T0310240

Charging for Mobile All-IP Telecommunications

Wiley Series on Wireless Communications and Mobile Computing

Series Editors: Dr Xuemin (Sherman) Shen, *University of Waterloo, Canada*
Dr Yi Pan, *Georgia State University, USA*

The "Wiley Series on Wireless Communications and Mobile Computing" is a series of comprehensive, practical and timely books on wireless communication and network systems. The series focuses on topics ranging from wireless communication and coding theory to wireless applications and pervasive computing. The books offer engineers and other technical professionals, researchers, educators, and advanced students in these fields with invaluable insight into the latest developments and cutting-edge research.

Other titles in the series:

Mišić and Mišić: *Wireless Personal Area Networks: Performance, Interconnections and Security with IEEE 802.15.4*, January 2007, 978-0-470-51847-2

Takagi and Walke: *Spectrum Requirement Planning in Wireless Communications: Model and Methodology for IMT-Advanced*, April 2008, 978-0-470-98647-9

Pérez-Fontán and Mariño Espiñeira: *Modeling the Wireless Propagation Channel: A Simulation Approach with MATLAB®*, August 2008, 978-0-470-72785-0

Ippolito: *Satellite Communications Systems Engineering: Atmospheric Effects, Satellite Link Design and System Performance*, September 2008, 978-0-470-72527-6

Myung: *Single Carrier FDMA: A New Air Interface for 3GPP Long Term Evolution*, November 2008, 978-0-470-72449-1

Hart, Tao and Zhou: *IEEE 802.16j Mobile Multihop Relay*, March 2009, 978-0-470-99399-6

Qian, Muller and Chen: *Security in Wireless Networks and Systems*, May 2009, 978-0-470-51212-8

Wang, Kondi, Luthra and Ci: *4G Wireless Video Communications*, May 2009, 978-0-470-77307-9

Cai, Shen and Mark: *Multimedia for Wireless Internet – Modeling and Analysis*, May 2009, 978-0-470-77065-8

Stojmenovic: *Wireless Sensor and Actuator Networks: Algorithms and Protocols for Scalable Coordination and Data Communication*, August 2009, 978-0-470-17082-3

Charging for Mobile All-IP Telecommunications

Yi-Bing Lin

National Chiao Tung University (NCTU), Taiwan

Sok-Ian Sou

National Cheng Kung University (NCKU), Taiwan

A John Wiley and Sons, Ltd, Publication

This edition first published 2008

Registered office
John Wiley & Sons Ltd, The Atrium, Southern Gate, Chichester, West Sussex, PO19 8SQ, United Kingdom

For details of our global editorial offices, for customer services and for information about how to apply for permission to reuse the copyright material in this book please see our website at www.wiley.com.

Library of Congress Cataloging-in-Publication Data

Lin, Yi-Bing.
Charging for mobile all-IP telecommunications / Yi-Bing Lin and Sok-Ian Sou.
p. cm. — (Wiley series on wireless communications and mobile computing)
Includes bibliographical references and Index.
ISBN 978-0-470-77565-3 (cloth)
1. Internet telephony—Prices. 2. Cellular telephone services industry.
3. Invoices—Computer programs. I. Sou, Sok-Ian. II. Title.
TK5105.8865L56 2008
384.5′33—dc22

2008013167

A catalogue record for this book is available from the British Library.

ISBN 978-0-470-77565-3 (HB)

Set in 11/13pt Times by Integra Software Services Pvt. Ltd, Pondicherry, India
Printed in Singapore by Markono Print Media Pte Ltd

This book is dedicated to my wife Sherry, my daughters Denise and Doris.

Yi-Bing Lin

Contents

List of Figures

List of Tables

Preface

Since Alexander Graham Bell and his assistant Thomas A. Watson built the first telephone prototype on March 10, 1876, telephone equipment and networks have been rapidly deployed worldwide. (Note that some say that the first telephone set, called "teletrofono", was invented and demonstrated by Antonio Meucci in early 1860.) The *Public Switched Telephone Network* (PSTN), which allows people to communicate with each other anywhere and anytime with natural voice, is considered to be the most complex system in the world. Based on the PSTN, the evolution of radio and mobile core network technologies over the last three decades has enabled the development of the ubiquitous mobile telecommunications. It provides mobile users with voice, data, and multimedia services at any time, any place, and in any format.

Alexander Graham Bell (Sketched by Yi-Bing Lin)

In order to charge for mobile telephony services, the telecom operator has to capture network usage in the form of charging records for all subscribers. These charging records are then processed to generate the subscriber's phone bill. Traditional telecom operators provide offline charging where the charging records are collected and sent to the billing system after the service is delivered. Traditional telecom operators also support prepaid services with limited billing functionalities. Advanced mobile telecom introduces data applications with real-time control and management, which requires a convergent and flexible online charging system for operators who want to mitigate fraud and credit risks. Online charging allows both prepaid and postpaid subscribers to be charged in real time. This feature is important in order to ensure that subscribers using telecom services do not make a purchase that would either exhaust their prepaid balance or exceed their postpaid credit limit.

In this book, we provide comprehensive and practical aspects of charging based on mobile operators' experiences and the latest efforts undertaken by the *Universal Mobile Telecommunications System* (UMTS) specifications. We present a complete overview of the telecom billing system, including the evolution from 2G/3G to *Next Generation Network* (NGN) all-IP network charging frameworks, and management aspects related to the charging and billing process. Specifically, this book describes the IP-based online charging system, protocol details and recent trends in charging for the mobile telecom industry.

We also include eight appendices in this book. Appendices A–F and H address some modeling issues for offline and online charging. Appendix F gives an example for application server implementation that supports online charging. Appendix G is a minor revision of Chapter 17 of the book *Wireless and Mobile Network Architectures* (Wiley 2001) by Lin and Chlamtac. This appendix provides an overview of the traditional mobile prepaid service approaches, which allows the readers to have better understanding of the distinctions between the modern prepaid solutions (which are the main focus of this book) and the traditional ones (as discussed in Appendix G).

Finally, we would like to thank our editor at Wiley. The writing of this book was supported in part by NSC 96-2752-E-009-005-PAE, NSC 96-2219-E-009-019, NSC 96-2221-E-009-020, Intel, Chung Hwa Telecom, IIS/Academia Sinica, ITRI/NCTU Joint Research Center and MoE ATU. The work of S.-I. Sou was supported by the Media Tek Fellowship.

Yi-Bing Lin & Sok-Ian Sou

1

Introduction

Niklas Zennström, a co-founder of Skype, stated: "It's the same thing with Skype. Some users are paying for services, but not everyone." Although services such as Skype have promoted free phone calls, *Charging, Billing and Accounting* (CBA) are still among the most important activities in telecommunications operation today. The charging function collects information related to chargeable events, correlates and processes the charging data, which are sent subsequently to the billing functional entity. The billing function processes the records coming from the charging functional entity according to the respective tariffs in order to calculate the charge for which the user should be billed. For example, it processes the call detail records to create some final output, which can be invoices for customers. The accounting function enables an automatic procedure for sharing of revenues, where the portion that is due to each player is calculated based on the agreement between the involved players [Kou04].

In mature markets, service differentiation is a key driver for telecom operators and service providers. As the convergence of services and networks accelerates, a major challenge will be to offer more services, innovative business models, and enhanced service accessibility. Charging and billing are a key part of the service delivery chain that supports the service differentiation challenge. They must offer the necessary flexibility and universality in order to follow, and sometimes anticipate, the evolution of service offer catalogs.

Various billing policies have been exercised in the *Public Switched Telephone Network* (PSTN). For example, in Taiwan, charges for local calls are on a per-minute basis, while charges for cellular services are on a per-minute basis or a per-second basis. Conversely, in the US, local calls can be provisioned under a monthly flat-rate charge. Flat-rate charging reduces the per-call billing cost, which is a primary motivation in the recent evolution of billing systems.

Charging for Mobile All-IP Telecommunications Yi-Bing Lin and Sok-Ian Sou

Billing World [Mor96] reported that two of the most desirable attributes of telecommunications billing systems are:

- the flexibility of upgrading it; and
- the capability to quickly inform the in-house billing experts about its status.

Telecommunications services are "culture sensitive" and the method of service charging significantly affects customer behavior. For example, since US cellular users are charged for radio access whether they place or receive calls, they tend to share their phone numbers carefully to avoid "junk calls". In Taiwan, however, since called parties are not billed, cellular users tend to distribute their phone numbers as widely as possible to enhance their business opportunities. To maximize profits, a cellular carrier needs to offer a variety of billing plans for the same services, and may need to adapt the plans to the "customer culture". Thus, the ease of upgrading is an important attribute of the billing system.

Another important attribute is the timely provision of billing status reports, which are essential for monitoring and diagnosing the billing system. One way to achieve this attribute is to report the customer billing records in real time. In PSTN, the real-time billing information is delivered through the *Signaling System No.7* (SS7) network. There are three formats for customer billing:

- The *full-detail format* provides all details (including the date, time elapsed prior to connection, calling number, called number, duration of call and charge) for each call.
- The *bulk-bill format* indicates the usage amount, the allowance amount, and the total usage charges over the allowance amount of a call.
- The *summary-bill format* provides the bulk-bill information and a summary showing total calls, overtime minutes, and the associated charge.

The full-detail format is typically used for long-distance calls while the bulk-bill and summary-bill formats are used for local calls. The billing information is produced from the *Automatic Message Accounting* (AMA) record, which is used primarily by local telephone companies to process billing records and exchange records between systems. The AMA record created by Telcordia (formerly Bellcore) is typically delivered in an SS7 message. During the call setup/release process, the monitor system tracks SS7 messages of the call and, once the call is complete, generates a *Call Detail Record* (CDR) in the AMA format. The CDR is then transferred to the rating and billing system.

In the traditional circuit-switched domain, a CDR is the computer record produced by a telephone exchange containing details of a call that passed through it. In the packet-switched domain, a CDR is generated by a network node for a data session. The "CDR" term is referred to as a *Charging Data Record*. In this book, CDR stands for "Call Detail Record" in a typical telephone call, and for "Charging Data Record" in a data session.

The possibility of cellular users roaming from their "home network" to a "visited network" causes difficulty in providing real-time customer billing records. When a cellular user is in a visited network, the billing records for all call activities remain in the visited network. In the 2G cellular roaming management/call control protocol [Lin95], the visited network and the home network do not interact at the end of a call. Instead, the billing information remains in the visited system as a "roam" type data record in *Cellular Intercarrier Billing Exchange Roamer Record* (CIBER) [Cib] or *Internet Protocol Detail Record* (IPDR) [Ipd] formats. These data records are batched and periodically sent to a clearinghouse electronically, and are forwarded by the clearinghouse to the customer's home network later. Alternatively, a roaming user manager may be responsible to forward billing records concerning roaming users to their home operators and support the *Transferred Account Procedure* (TAP), which converts and groups billing records in files under the TAP format in order to be sent to the respective operators.

The performance of the transmission of billing information depends on the frequency of information exchange. In an ideal case, records would be transmitted for every phone call to achieve real-time operation. However, real-time transmission would significantly increase cellular signaling traffic and seriously overload the PSTN signaling network. In order to achieve billing status reports quickly, a tradeoff should be made to balance the frequency of the billing transmission with the signaling traffic [Fan99].

In the *Universal Mobile Telecommunications System* (UMTS), a *Charging Gateway* (CG) can generate CDRs and transfer them to the billing system using the *File Transfer Protocol* (FTP) [IET85]. The CDRs are transferred in either push or pull modes [3GP06b]. In the push mode, the CG pushes the charging records to the billing system. CDRs generated from the charging node are transferred to the billing system at a frequency controlled by the CG. In the pull mode, the billing system pulls the charging records from the CG. The time and frequency of the file transfer is controlled by the billing system.

1.1 Charging for Mobile All-IP Networks

By providing ubiquitous connectivity for data communications, the Internet has become the most important vehicle for global information delivery. Furthermore, introduction of the 3G mobile system has driven the Internet into new markets to support mobile users. Specifically, the *Internet Protocol* (IP) has played a major role in UMTS in providing a wide range of connectionless services to mobile users [Lin05a].

The Internet environment encourages global usage with flat-rate tariffs and low entry costs. A major problem of the "flat-rate" tariffs is that such a business model cannot justify the expensive equipment/operation investments of mobile services [Das00,

Fal00, Rei06]. Mobile telecom operators have to move from a bit-pipe model to a revenue-generating services model. To integrate IP with wireless technologies with the "right" business model, the *3rd Generation Partnership Project* (3GPP) proposed UMTS all-IP architecture to enable web-like services and new billing paradigm in the telephony world. This architecture has evolved from *Global System for Mobile Communications* (GSM), *General Packet Radio Service* (GPRS), and UMTS Release 1999 to Releases 4–8. UMTS Release 4 introduces the *Next Generation Network* (NGN) architecture for the *Circuit Switched* (CS) and the *Packet Switched* (PS) service domains. UMTS Release 5 introduces the *IP Multimedia core network Subsystem* (IMS) on top of the PS service domain. This evolution requires new mechanisms to collect information about chargeable events and to impose flexible mobile billing schemes (such as time-based, volume-based, or content-based).

A telecom operator typically provides *offline charging* (referred to as postpaid charging) where the charging records are collected and then sent to the billing system once the service has been delivered. On the other hand, prepaid telecommunications service requires a user to make an advanced payment before the service is delivered. Usage of prepaid service does not require deposit and monthly bill. Instead, the usage fee is directly deducted from the user's prepaid account. (Figure 1.1 shows an example of prepaid recharge webpage. The users can refill the prepaid accounts through this

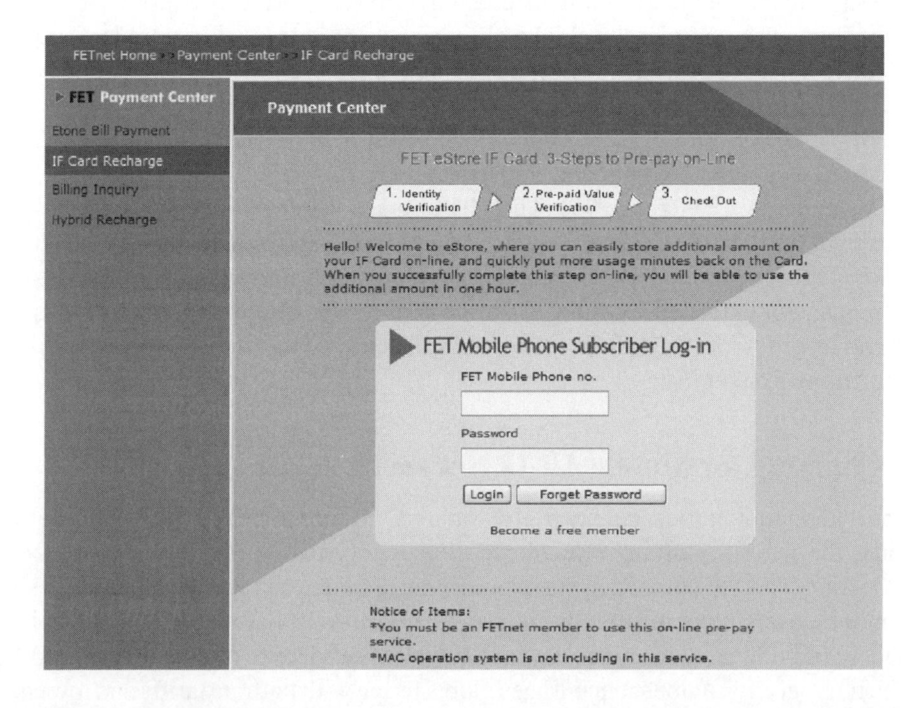

Figure 1.1 A recharge webpage for prepaid card (reproduced by permission of Far EasTone Telecommunications Co., Ltd)

webpage when the prepaid credits are completely or nearly depleted.) The prepaid viewpoint for telecommunications services has been deployed and studied for more than a decade (see Chapter 17 in [Lin01]). This service was offered in Europe and Asia in 1982 and became popular in the US in 1992 [Art98]. In Europe and quite a few Asian markets, more than 50% of new mobile subscribers are prepaid users [HP02]. Four billing technologies have been used in traditional (circuit-switched) mobile prepaid service: hot billing approach [Lin00]; service node approach [Cha02a, Cha02b]; handset-based approach; and *Intelligent Network* (IN) approach [Lin06]. Details of these approaches will be given in Appendix G. The first three approaches are already outdated due to significant operation and maintenance costs. The fourth approach still exists and has been evolved from early IN versions to the *Customized Application for the Mobile Network Enhanced Logic* (CAMEL) approach to be described in Chapter 5.

A traditional prepaid approach is typically used to handle only one voice call session at a time. In this case, all prepaid credits are allocated to the in-progress circuit-switched voice session. In advanced mobile data services, the data sessions are packet switched, and multiple prepaid sessions must be accommodated simultaneously. To address the above issues, the telecom networks have evolved from the SS7-based architecture into the all-IP based architecture [3GP, DIA, Kur06a, Kur06b]. The main focus of this book is on pure IP-based mobile charging techniques.

Advanced mobile telecom incorporates data applications with real-time control and management, which requires a convergent and flexible online charging system. Such convergence is essential to mitigate fraud and credit risks and provide more personalized advice to users about charges and credit limit controls. *Online charging* allows simultaneous prepaid and postpaid sessions to be charged in real time. This feature is important for telecom operators to support multiple service deliveries simultaneously. The subscribers cannot make purchases that either exhaust the prepaid balance or exceed the credit limit. Through online charging, the operator can ensure that credit limits are enforced and resources are authorized on a per-transaction basis. From a subscriber's perspective, knowing the charges in advance and having self-imposed credit limits can make the cost of services more transparent. Real-time rating/charging functions help subscribers to control their budgets, and telecom operators to reduce bad debts.

By merging the prepaid and postpaid methods, the *Online Charging System* (OCS) [3GP06a] proposed in UMTS Releases 5 and 6 eliminates duplicate processes such as service development, production support, pricing, and general management activities. In other words, online charging only requires a single set of management processes, which reduces operational costs and enhances flexibility on billing and product diversification. The OCS approach is estimated to reduce 25% of the product launch costs [Con05]. This real-time solution provides two-way communications between network nodes and the charging/billing system, which transfer information about rating, billing and accounting. When the OCS receives a service request from an application server (or a network node), it queries other relevant components, then determines and returns a response to the application server. In contrast, in an offline environment, all usage records typically flow through the billing system in one direction after the service has been delivered.

The IMS architecture supports both online and offline charging methods. In online charging, IMS nodes dynamically interact with the OCS. The OCS in turn processes the user's account in real time and controls the service charges. In the IMS offline method, the IMS network nodes collect the charging information and then send to the billing system after the service is delivered. The charging/billing system does not control the service usage in real time. In this model a user typically receives the bill on a monthly basis.

1.2 Online Charging

The OCS is flexible enough to support multiple charging/billing models. For example, a mobile user might be charged by connection time or by the amount of data transferred. The user may also be given free access to certain websites. Some subscribers may use prepaid billing and others may use postpaid billing. The OCS can support all these requirements for a service provider. Consider a bundle mixing triple play and mobile services for a family, which requires highly flexible rating and community charging capabilities offered by the OCS [Alc07]. The package consists of the following items:

- A 60-euro monthly fee, including triple play services covering unlimited national calls (VoIP), Internet access, and access to selected TV channels (IPTV); a mobile bundle with two hours of mobile communications to national mobile numbers, 30 national short messages, and 10 national multimedia messages.
- The children have their own mobile prepaid account limited to 30 minutes of national voice and 10 short messages debited from the family bundle. They can top up their prepaid account with scratch cards when the balance is exhausted.
- All other calls (outside the monthly fee) are charged by time and destination.
- All services are charged on a single monthly bill.

From the OCS rating viewpoint, the following functions are required:

- Management of the monthly fee (60 euros) with several balances in units (short messages, minutes, multimedia messages, and so on).
- Community management, with routing of charges to the family account, and a limit on the children's usage.
- Out-of-bundle call charging according to the appropriate tariff plan.
- CDR generation and aggregation for the monthly bill.

When a subscriber is first authorized and starts to access the service, the network nodes may send periodic credit control messages to the OCS. For example, the IMS application server might send a credit control request for every megabyte of data downloaded. When the OCS receives these credit control message requests, it updates the subscriber's account accordingly.

The OCS supports *prepaid service* that requires a user to make a payment in advance before enjoying the service. Prepaid service does not require a monthly bill. Instead, the user's prepaid account directly deducts the fee. Depending on the type of requested prepaid service, the OCS reserves an amount of prepaid quota in the subscriber's account before authorizing the service. The OCS deducts the prepaid credit in real time when the prepaid service is delivered.

An example of a prepaid service is described as follows: Suppose that a prepaid subscriber wants to download an MP3 music file that costs $5.00. The IMS application server first sends the reservation request to the OCS. The OCS then determines the price and reserves $5.00 in the subscriber's account. The OCS authorizes the transaction by sending a message to the application server. When the subscriber completes downloading the music file, the application server sends an accounting message to the OCS to indicate that the service has been delivered. The OCS then deducts the reserved $5.00 from the subscriber's account. If the downloading fails (because, for instance, the subscriber cancels the request), the application server sends an accounting message to indicate that the service was not delivered. The OCS then reclaims the reserved $5.00 for the subscriber's account. If the subscriber attempts to download the MP3 file again, the service must be re-authorized.

Another example for online charging is described as follows: Suppose that a subscriber pays for access based on the duration of the connection; for instance, the fee for playing 10 minutes of an online mobile game might be $1.00. Assume that a subscriber has $10.00 in her prepaid account. When she attempts to play the online mobile game, the OCS authorizes and reserves $1.00 for the first 10 minutes of connected time. The subscriber starts playing the mobile game; after 10 minutes, the IMS application server informs the OCS that the subscriber has been playing for 10 minutes. The OCS then deducts $1.00 from the subscriber's account. If the subscriber continues to play the game, the OCS reserves another $1.00 for every additional 10 minutes of connected time.

After the subscriber has played for 100 minutes, her account will be depleted. The next reservation request will be denied. In this situation, the application server might inform the subscriber that she needs to replenish her account before continuing to play the game. The subscriber might purchase a refill card or replenish the account by using her credit card.

The OCS can be configured to determine the status of each subscriber based on the account type (prepaid or postpaid), the duration of service (long-time or new subscribers), or amount of money spent in the last month (high-usage or low-usage subscribers). Different types of subscribers could be handled differently. For example, long-time subscribers could be given preferential treatment. New or high-usage subscribers could be given free access time for promoted services.

The OCS also supports price enquiry; e.g., through *Advice of Charge* (AoC). If a subscriber wants to know the estimated price of the IMS service before purchasing it, the IMS application server issues a price enquiry to the OCS, which checks the fee and returns the estimated price to the application server. The application server then shows the estimated price to the subscriber.

1.3 Concluding Remarks

In this chapter, we introduced telecom charging and billing, which are two of the most important activities in telecommunications networks. We described various billing methods exercised in the PSTN. We elaborated on the offline and online charging mechanisms for mobile all-IP networks and described examples for online charging. Throughout this book, we will introduce advanced charging techniques and mechanisms. Here we emphasize that service providers need to use these charging models wisely so as not to infringe on their ethical responsibility to their customers. As Judge Harold H. Greene said, "The emphasis on getting business and billing goes far beyond the professionalism of law and does the profession injustice." Such injustice is what we must avoid.

Review Questions

1. Describe two of the most desirable attributes for telecommunications billing systems.
2. Which protocol can be used for real-time billing transmission in the PSTN?
3. Describe three formats for customer billing. How are these formats used?
4. What is the difficulty of providing real-time billing records in a cellular system?
5. Which protocol is used for CDR transmission to a billing system in UMTS?
6. Describe two billing record formats used for roaming in a cellular system.
7. Briefly describe the evolution of the UMTS all-IP network.
8. Describe how the traditional prepaid mechanisms (such as hot billing approach, service node approach, and handset-based approach) work. Why are they outdated? (Hint: see Appendix G.)
9. Describe online charging and offline charging. Why is online charging important?
10. Show two online charging services and briefly describe how they work.

References

[3GP] The 3rd Generation Partnership Project (3GPP) (http://www.3gpp.org).

[3GP06a] 3GPP, 3rd Generation Partnership Project; Technical Specification Group Service and System Aspects; Telecommunication management; Charging management; Online Charging System (OCS): Applications and interfaces (Release 6), 3G TS 32.296 version 6.3.0 (2006-09), 2006.

[3GP06b] 3GPP, 3rd Generation Partnership Project; Technical Specification Group Service and System Aspects; Telecommunication management; Charging management; Charging Data Record (CDR) file format and transfer (Release 6), 3G TS 32.297 version 6.2.0 (2006-09), 2006.

[Alc07] Alcatel-lucent 8610 Instant Convergent Charging Suit: Fixed Mobile Convergence Overview, Release 4.5, 2007.

[Art98] Arteta, A., Prepaid billing technologies – Which one is for you? *Billing World*, February, pp. 54–61, 1998.

[Cha02a] Chang, M.-F. and Yang, W.-Z., Performance of mobile prepaid and priority call services, *IEEE Communications Letters*, **6**(2): 61–63, 2002.

[Cha02b] Chang, M.-F., Yang, W.-Z., and Lin, Y.-B., Performance of hot billing mobile prepaid service, *IEEE Transactions on Vehicular Technology*, **51**(3): 597–612, 2002.

[Cib] http://www.cibernet.com/Clearing/CIBER.htm.

[Con05] Convergys, Online charging: Delivering pre- and post-paid convergence, Convergys White Paper, 2005.

[Das00] DaSilva, L.A., Pricing for QoS-enabled networks: a survey, *IEEE Communications Surveys*, Second Quarter 2000.

[DIA] EU FP6 IST IP Daidalos and Daidalos II (http://www.ist-daidalos.org).

[Fal00] Falkner, M., *et al.*, An overview of pricing concepts for broadband IP networks, *IEEE Communications Surveys*, Second Quarter 2000.

[Fan99] Fang, Y., Chlamtac, I., and Lin, Y.-B., Billing strategies and performance analysis for PCS networks, *IEEE Transactions on Vehicular Technology*, **48**(2): 638–651, 1999.

[HP02] HP. Generating revenue with a new breed of prepaid user, HP White Paper, 2002.

[IET85] IETF. File Transfer Protocol (FTP). IETF RFC 959, 1985.

[Ipd] http://www.ipdr.org/.

[Kou04] Koutsopoulou, M. and Kaloxylos, A., A holistic solution for charging, billing and accounting in 4G mobile systems, *Vehicular Technology Conference* (VTC), **4**: 2257–2260, May 2004.

[Kur06a] Kurtansky, P. and Stiller, B., State of the art prepaid charging for IP services. Conference on Wired/Wireless Internet Communications (WWIC). *LNCS*, Springer, **3970**:143–154, May 2006.

[Kur06b] Kurtansky, P., *et al.*, Efficient prepaid charging for the 3GPP IP Multimedia Subsystem (IMS). *SDPS*, **1**:462–472, San Diego, USA, June 2006.

[Lin95] Lin, Y.-B. and DeVries, S., PCS network signaling using SS7. *IEEE Personal Communications Magazine*, **2**(3): 44–55, 1995.

[Lin00] Chang, M.-F., Lin, Y.-B., and Yang, W.-Z., Performance of hot billing mobile prepaid service, *Computer Networks Journal*, **36**(2): 269–290, 2001.

[Lin01] Lin, Y.-B. and Chlamtac, I., *Wireless and Mobile Network Architectures*. John Wiley & Sons, Ltd., Chichester, UK, 2001.

[Lin05a] Lin, Y.-B., and Pang, A.-C., *Wireless and Mobile All-IP Networks*. John Wiley & Sons, Ltd., Chichester, UK, 2005.

[Lin06] Lin, P., Lin, Y.-B., Yen, C.-S., and Jeng, J.-Y., Credit allocation for UMTS prepaid service, *IEEE Transactions on Vehicular Technology*, **55**(1): 306–317, 2006.

[Mor96] Moran, S.H., Cellular companies to expend up to $1.2 billion on billing and customer care in 1996, *Billing World*, pp. 12–16, March/April, 1996.

[Rei06] Reichl, P., *et al.*, A stimulus-response mechanism for charging enhanced quality-of-user experience in next generation all-IP networks, *13th Latin-Ibero American Operations Research Conference* (CLAIO 2006), Montevideo, Uruguay, November 2006.

2

Telecommunications Networks

When the first telephone machine was invented in 1876, there were many arguments about its usage. A Western Union internal memo stated: "Telephone has too many shortcomings to be seriously considered as a means of communication. The device is inherently of no value to us." Fortunately, the above criticism has been proven wrong. Today, many telecommunications networks have been deployed to link people from all parts of the globe.

This chapter provides an overview about telecommunications networks. We first introduce the traditional *Public Switched Telephone Network* (PSTN) and describe the *Universal Mobile Telecommunications System* (UMTS) that integrates the Internet with wireless technologies. We then elaborate on the *IP Multimedia core network Subsystem* (IMS) developed based on UMTS, which enhances the functionality of the packet-switched service domain by supporting IP-based multimedia applications. Finally, we describe 3GPP *Wireless Local Area Network* (WLAN) that can interwork with UMTS.

2.1 Public Switched Telephone Network

In the PSTN, voice, data and other telecommunications services are carried over dedicated transmission paths. A path is established using normal telephone signaling and ordinary switched telephone circuits. A subscriber uses *Customer Premises Equipment* (CPE; Figure 2.1(a)), a set of telecommunications equipment located at a customer's premises (except for pay phones), to communicate with other parties. CPE can be either voice grade analog equipment, which is suitable for transmitting a voice signal from 300 Hz to 3400 Hz, or data grade digital equipment, which is suitable for transmitting high frequency signals such as 4 Mbps. A data grade CPE generates digital pulses that must be converted to analog signals before they are delivered to the PSTN. Fax machines, ISDN phones and *Mobile Stations* (MSs) – which are handsets for

Charging for Mobile All-IP Telecommunications Yi-Bing Lin and Sok-Ian Sou

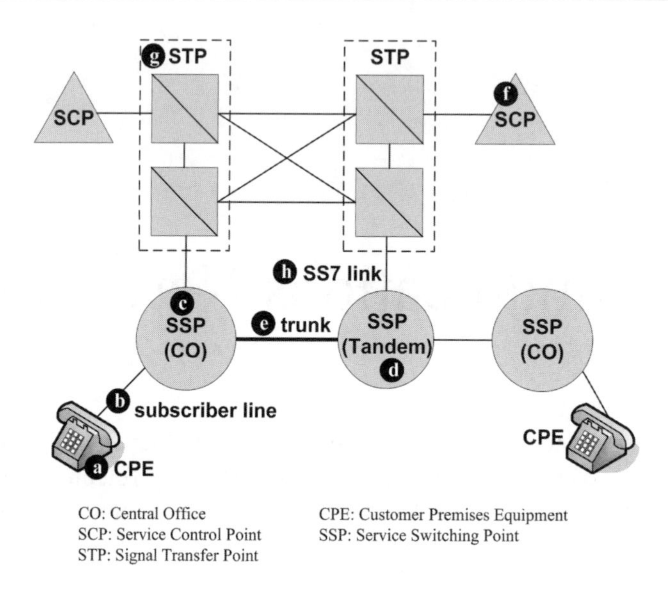

Figure 2.1 The PSTN architecture

accessing cellular service – are all examples of CPEs. Another type of CPE is *Private Branch Exchange* (PBX; see Figure 2.2) installed in a corporation. A PBX is a small local telephone office that provides telephone extensions for business use and access to the PSTN. Not only does the PBX enable intelligent transmission and information processing, it also has its own numbering plan: a typical PBX extension is made up of three to six digits. Internal calls are dialed directly to an extension number, while outside calls are first answered by the PBX attendant, and then transferred to the corresponding extension.

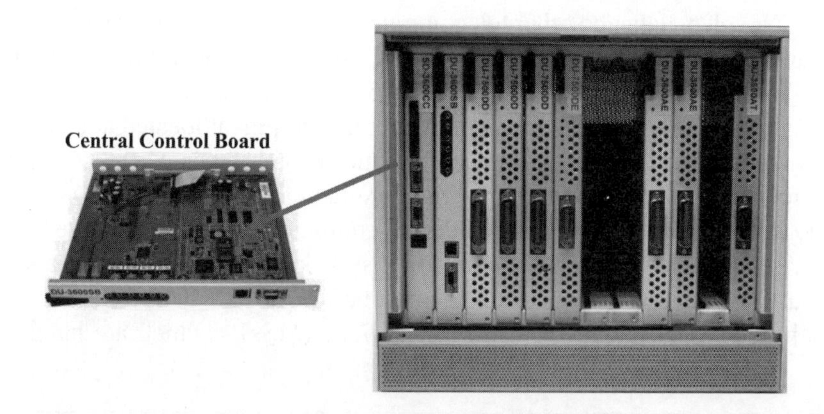

Figure 2.2 IP-based PBX system (reproduced by permission of TECOM Co. Ltd)

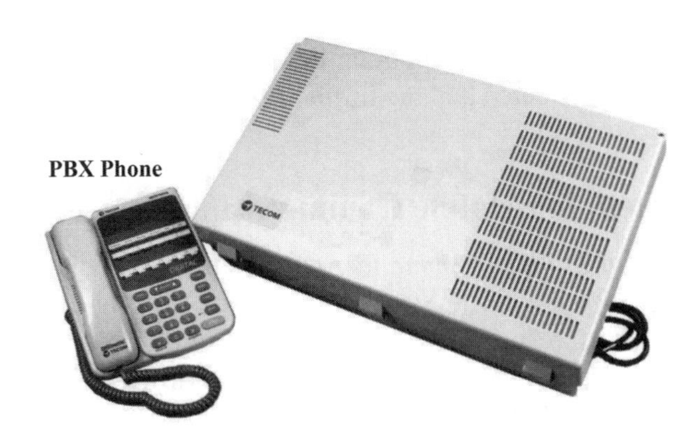

Figure 2.2 (*continued*)

The CPE connects to a *Central Office* (CO; Figure 2.1(c)) through the subscriber line (Figure 2.1(b)). A CO is a public switching system that connects to the CPE, while a *tandem* (Figure 2.1(d)) is a switching system that connects to COs or other tandems through *trunks* (Figure 2.1(e)). As a general rule, a tandem does not directly connect to the CPE. A local tandem connects to local COs in the same charging zones (for local calls); a toll tandem, or interexchange tandem, connects COs in different charging zones (for long distance calls). Although a PBX is usually considered as a CPE, it shares the same basic switch architecture as that of the public switching systems.

In the PSTN, call setup/release and enhanced services are controlled by signaling systems such as the *Signaling System No. 7* (SS7). The SS7 signaling network consists of three distinct components [Lin95, Lin01].

- A *Service Switching Point* (SSP) is a telephone switch such as a CO or a tandem (Figure 2.1(c) and (d)) that processes calls which originate, tandem, or terminate at that node.
- A *Service Control Point* (SCP; Figure 2.1(f)) contains databases to provide enhanced services. The SCP accepts queries from an SSP and returns the requested information to the SSP.
- A *Signal Transfer Point* (STP; Figure 2.1(g)) is a switch/router that relays SS7 messages between SSPs and SCPs. To meet the stringent reliability requirements, STPs are provisioned in mated pairs.

Digital bi-directional signaling links (Figure 2.1(h)) connect the SS7 components. Logically, SCPs and the SSPs are indirectly connected through the STPs. To support high availability, the SS7 networks are deployed with multiple links among the network components. The STPs are configured in pairs and databases/applications are duplicated in different SCPs. The signaling links are monitored such that the failure links are automatically detected and the traffic load is shared by the active links.

2.2 Global System for Mobile Communications

The *Global System for Mobile Communications* (GSM) is a digital wireless network specified by standardization committees from major European telecommunications operators and manufacturers. The GSM standard provides a common set of compatible services and capabilities to all mobile users worldwide. Figure 2.3 illustrates the GSM architecture: a *Mobile Station* (MS; Figure 2.3(a)) communicates with the network through the *Base Station System* (BSS; Figure 2.3(b)) that consists of the *Base Transceiver Station* (BTS; Figure 2.3(c)) and the *Base Station Controller* (BSC; Figure 2.3(d)). The BTS communicates with the MS via the radio interface while the BSC communicates with the *Mobile Switching Center* (MSC; Figure 2.3(e)) via the A interface. An MSC is a special telephone switch (i.e., SSP) that supports mobile applications and creates charging records to the billing system (Figure 2.3(i)). An MSC supports switching and mobility management functions and connects with other MSCs and/or SSPs in the PSTN (Figure 2.3(f)).

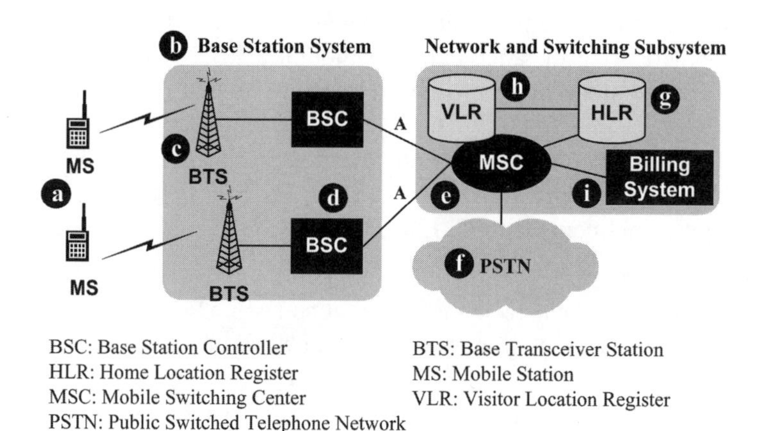

Figure 2.3 The GSM network architecture

When a user subscribes to the GSM services, a record is created in the system's database called *Home Location Register* (HLR; Figure 2.3(g)). This HLR record contains several fields including the *Mobile Station ISDN Number* (*MSISDN*) and the *International Mobile Subscriber Identity* (IMSI). The MSISDN is a telephone number of the MS; the IMSI identifies the MS in the GSM network. Unlike the MSISDN, the IMSI is not known to the users, and is used in the network only [Lin01]. The IMSI, whose length is no more than 15 digits, consists of three parts: a three-digit

Mobile Country Code (MCC); a two- or three-digit *Mobile Network Code* (MNC); and a *Mobile Station Identification Number* (MSIN).

The GSM service area is partitioned into several registration areas, which each consists of a group of BTSs. When a mobile user enters a new registration area, the *Visitor Location Register* (VLR; Figure 2.3(h)) creates a temporary record for the MS. The VLR then informs the MS's HLR of its current location. The HLR and the VLRs provide subscriber profiles and mobility management, and are typically implemented in the SCP platform. A *Gateway MSC* (GMSC) is capable of querying the HLR to determine the current location of an MS (i.e., the SS7 address of the visited MSC). Therefore, the GMSC can then route the calls to the MS through the visited MSC. We will elaborate on the details of GSM call setup in Section 6.1.

GSM supports *Short Message Service* (SMS), which allows mobile users to send and receive simple text messages up to 140 octets. The GSM SMS network architecture is illustrated in Figure 2.4. In this network, an originating MS sends a short message to the BTS/BSC. The BSC then forwards the message to an MSC called SMS *Inter-Working MSC* (IWMSC; Figure 2.4(a)). The IWMSC passes this message to a *Short Message Service Center* (SM-SC; Figure 2.4(b)). Upon receipt of the message, the SM-SC forwards it to the destination GSM network through a specific MSC called the *SMS Gateway MSC* (SMS GMSC; Figure 2.4(c)). Following the GSM roaming protocol, the SMS GMSC locates the target MSC of the message receiver and forwards the message to this MSC. The target MSC sends the message to the terminating MS through the BSC/BTS. After SMS delivery, the MSCs provide the charging records to the billing system (Figure 2.4(d)).

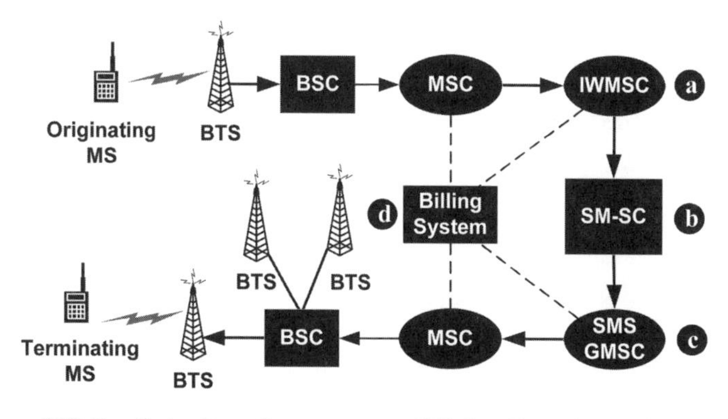

BSC: Base Station Controller
BTS: Base Transceiver Station
IWMSC: Inter-Working MSC
MS: Mobile Station
MSC: Mobile Switching Center
SM-SC: Short Message Service Center
SMS GMSC: Short Message Service Gateway MSC

Figure 2.4 GSM short message service architecture

2.3 Universal Mobile Telecommunications System

In early 2000, only a small portion of GSM subscribers used data services because existing GSM data transmission services was too expensive. To offer better mobile data services, GSM operators introduced the *General Packet Radio Service* (GPRS), which reuses the GSM infrastructure to provide end-to-end packet-switched services. In GPRS, the existing GSM nodes such as BSS, MSC, VLR and HLR are upgraded to improve the efficiency of packet data transmission.

To support high-speed packet data for versatile multimedia services, UMTS has evolved from GPRS by mainly replacing the radio access network. Figure 2.5 shows the architecture for the UMTS *Packet-Switched* (PS) service domain. In this figure, the dashed lines represent signaling links and the solid lines represent data and signaling links. The PS *core network* is an IP-based backbone network that consists of *GPRS Support Nodes* (GSNs) such as *Serving GSNs* (SGSNs; see Figure 2.5(d)) and *Gateway GSNs* (GGSNs; see Figure 2.5(e)).

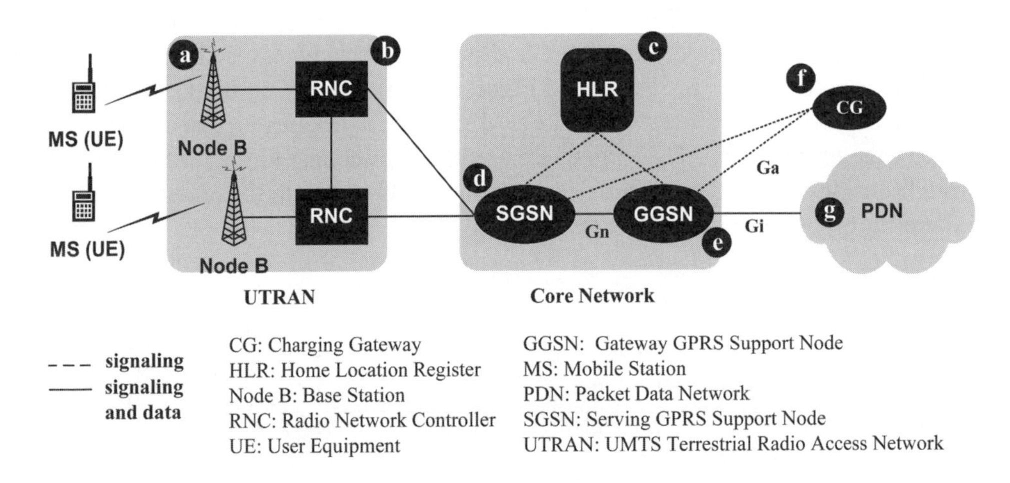

Figure 2.5 The UMTS network architecture

In the PS service domain, an SGSN connecting to the *UMTS Terrestrial Radio Access Network* (UTRAN) plays a role that is similar to the MSC in the circuit-switched (CS) service domain. The GGSN interworks the GPRS network with the external *Packet Data Network* (PDN; see Figure 2.5(g)). The HLR (Figure 2.5(c)) communicates with the GSNs for mobility and session management. The UTRAN consists of *Node B*s (the UMTS term for base stations; see Figure 2.5(a)) and *Radio Network Controller*s (RNCs; see Figure 2.5(b)) connected by an ATM or a high-speed network. An MS communicates with one or more Node Bs through the radio interface *Uu* based on the *Wideband CDMA* (WCDMA) radio technology [Che07, Hol04]. The

Charging Gateway (CG; see Figure 2.5(f)) collects the billing and charging information from the GSNs. Note that in the UMTS, an MS is also called a *User Equipment* (UE). Throughout this book, we will use the terms MS and UE interchangeably.

The 3GPP standard defines several IP-based interfaces among the GSNs, the CGs and the external PDN. In the Gn interface, the *GPRS Tunneling Protocol* (GTP) [3GP07] transports user data and control signals among the GSNs. Specifically, the GTP for the user plane (GTP-U) tunnels user data packets between GSNs; the GTP for the control plane (GTP-C) tunnels signaling messages between SGSNs and GGSNs. The control plane messages are used to create, update and delete GTP tunnels and for path management. The GGSN connects to the PDN through the Gi interface (which follows the Internet protocol). In the Ga interface, the GTP' protocol is utilized to transfer the *Charging Data Records* (CDRs) from GSNs to CGs (to be elaborated in Chapter 4). In UMTS, the CDRs are required for a number of telecom management activities, such as billing to home subscribers, settling accounts for visiting subscribers with other operators, analyzing service usage for statistics, and so on.

2.3.1 Packet Data Protocol (PDP) Contexts

Packet Data Protocol (PDP) contexts [Lin05] are created when a subscriber activates a GPRS session. These PDP contexts, which provide information to support packet delivery between an MS and the network, are stored in the MS, the SGSN and the GGSN. For each GPRS session of an MS, a PDP context is created to characterize the session. A PDP context contains the following information:

- *PDP Route Information* includes the PDP type, the PDP address and the GTP tunnel endpoint identifier. In UMTS, the PDP type is IP in general.
- *APN Information* indicates the external network to be accessed by the MS. An *Access Point Name* (APN) label represents access service to the Internet, the company intranet, or the WAP service to be described in Section 2.3.2.
- *QoS Information* includes QoS profile subscribed, QoS profile requested and QoS profile negotiated. After the subscription time, a user indicates the subscribed QoS for a specific service (e.g., the bandwidth for VoIP). When the user accesses this service, she specifies the requested QoS. Based on the available network resources, the GGSN determines the negotiated QoS for the service.
- *Charging Information* includes the charging characteristics such as normal, prepaid, flat-rate, or hot billing.

The charging characteristics in the PDP context are supplied by the HLR to the SGSN as a part of the subscription information during location update. Upon activation of a PDP context, the SGSN forwards the charging characteristics to the GGSN. The GSNs handle the GPRS session according to the charging characteristics methods defined in [3GP05a].

The message flow for PDP context activation procedure is shown in Figure 2.6, and occurs in the following sequence:

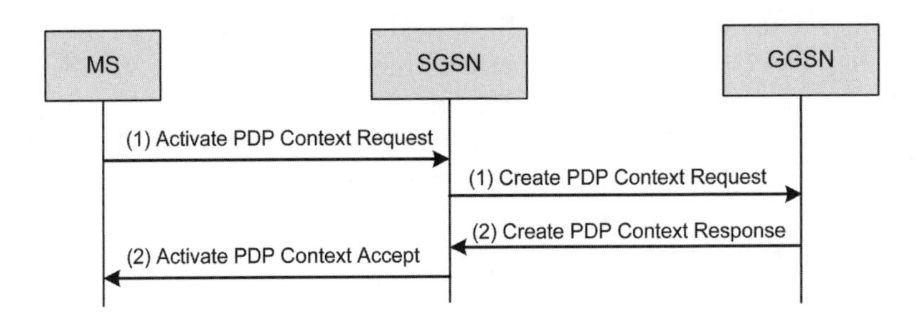

Figure 2.6 PDP context activation

Step 1. The MS sends the Activate PDP Context Request message with the PDP type, the PDP address (optional), the APN, and the requested QoS profile to the SGSN. Based on this message, the SGSN creates a PDP context. The SGSN then sends the Create PDP Context Request message to the GGSN. If the MS is not assigned a static IP address, then the GGSN is responsible for allocating a dynamic IP address to the MS.

Step 2. The GGSN uses the APN to find an external network that provides the requested service. In addition, the GGSN may restrict the negotiated QoS profile given its capabilities and the current load. The GGSN generates a *GPRS Charging Identifier* (GCID) and creates a PDP context associated with the MS. This PDP context records the requested APN, negotiated QoS profile, the GCID and so on. The GGSN then returns the Create PDP Context Response message to the SGSN to indicate that a PDP context has been created in the GGSN. This message provides the dynamic IP address, negotiated QoS and so on, which are filled in the PDP context of the SGSN. Through the Activate PDP Context Accept message (with the information such as the IP address assigned to the MS), the SGSN informs the MS that the GPRS session is set up.

After the PDP context activation, the MS can start the GPRS session with the external network. The PDP contexts are removed when the subscriber terminates the GPRS session. The message flow of the PDP context deactivation procedure appears in Figure 2.7, and occurs in the following steps:

Step 1. The MS sends the Deactivate PDP Context Request message to the SGSN. The SGSN sends the Delete PDP Context Request message to the GGSN.

Step 2. The GGSN removes the PDP context and returns the Delete PDP Context Response message to the SGSN. Then the SGSN deactivates (or removes) the PDP context, and returns the Deactivate PDP Context Accept message to the MS.

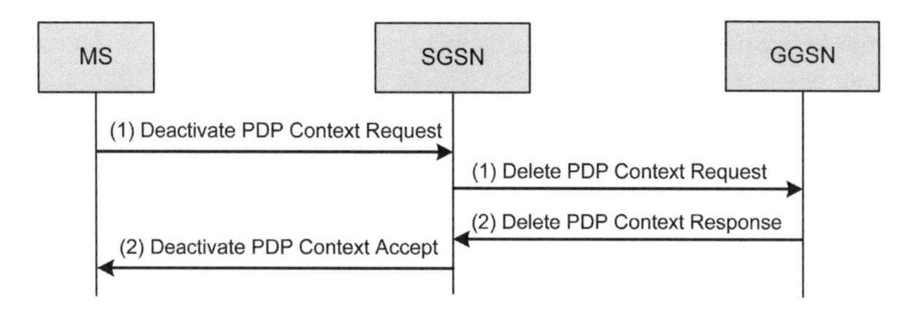

Figure 2.7 PDP context deactivation

In a GPRS session, the charging records are generated by the SGSN and the GGSN. These records are sent to the CG (Figure 2.5(f)) via the GTP' protocol described in Chapter 4. Details of GGSN CDR generation will be given in Section 6.2.

2.3.2 Wireless Application Protocol (WAP)

Several wireless data services such as *Wireless Application Protocol* (WAP) and *Multimedia Messaging Service* (MMS) can be delivered in the UMTS network. In mobile data networks, the data bandwidth is limited and the wireless handsets are constrained by small displays, powerless CPUs, limited memory, and awkward interfaces. As a solution to these issues, WAP presents and delivers wireless data services to mobile phones and other wireless terminals [Wap]. In particular, WAP facilitates the implementation of contents and applications on various types of wireless terminals. The WAP architecture is shown in Figure 2.8. In this figure, a WAP handset communicates with the origin server on the Internet through the UMTS network and the WAP gateway:

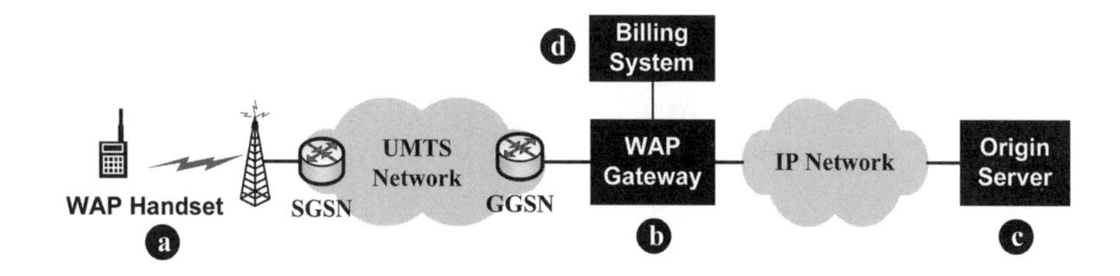

Figure 2.8 The WAP architecture

- The WAP handset (a UMTS MS with the WAP browser; see Figure 2.8(a)) receives content that is encoded in the compact binary format of *Wireless Markup Language* (WML).
- The WAP gateway (Figure 2.8(b)), located between the IP network and the mobile network, receives the WAP request from the handset, decodes the request from binary format into text format, and forwards it to the origin server. The

encoding/decoding operations significantly reduce the amount of data delivered in the low bandwidth radio link.
- The origin server (Figure 2.8(c)) is a standard *Hypertext Transfer Protocol* (HTTP) web server that contains certain resources (i.e., contents to be viewed by the users) and determines which content to be retrieved by parsing the request received from the WAP gateway.

When the WAP handset requests resources from an origin server, the request is sent to a WAP gateway and then to the origin server. The origin server retrieves and returns the requested resources to the WAP gateway which is responsible for encoding the content into binary codes. Then the WAP gateway transmits the encoded content to the WAP handset through the UMTS network. The user accesses the information by a micro-browser on the WAP handset. After service delivery, the WAP gateway sends the charging record to the billing system (Figure 2.8(d)).

Three kinds of messages are defined in WAP:

- The WAP PUSH message is a specially formatted SMS message that displays an alert message allowing the user to connect to a particular *Uniform Resource Locator* (URL) via the mobile phone's WAP browser.
- The WAP POST message is used to deliver data.
- The WAP GET message is used to retrieve data.

2.3.3 Multimedia Messaging Service (MMS)

In UMTS, MMS is introduced to deliver messages that range in size from 30K bytes to 100K bytes [3GP05b, 3GP06b]. Formats that can be embedded within MMS include text (formatted with fonts, colors, and other style elements), images (in JPEG or GIF formats), audio (in MP3 or MIDI formats), and video (in MPEG format). Figure 2.9 illustrates the MMS architecture that supports MMS delivery to a user agent (Figure 2.9(a)) that is built in an MS or an external device connected to the MS. Either the MMS *Value-Added Service* (VAS) applications (Figure 2.9(b)) which are

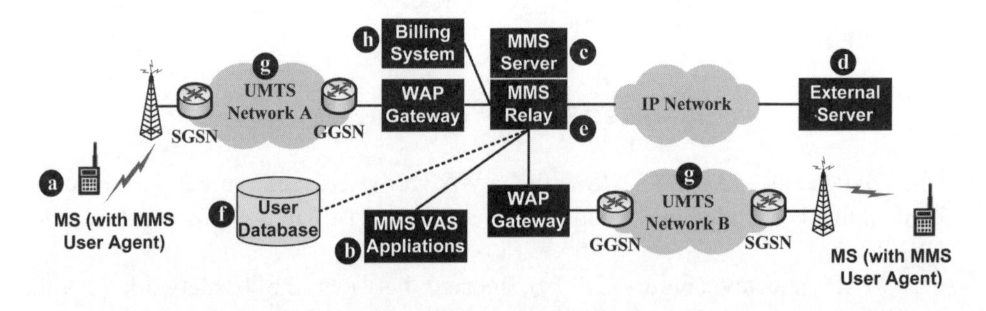

Figure 2.9 Multimedia messaging service architecture

connected to the mobile networks, or external servers (such as email server or fax server; see Figure 2.9(d)) in the IP network, provide the MMS contents. The *MMS Server* (Figure 2.9(c)) processes and stores incoming and outgoing multimedia messages; the *MMS Relay* (Figure 2.9(e)) transfers messages between different messaging systems, and adapts messages according to the capabilities of the receiving devices. It also generates charging data to the billing system (Figure 2.9(h)). The *MMS user database* (Figure 2.9(f)) contains user subscriber data and the configuration. The Internet protocol provides connectivity between different mobile networks (Figure 2.9(g)), which can be WAP-based 2.5G/3G networks illustrated in Figure 2.8. An MS can easily send out an MMS picture with the steps shown in Figure 2.10.

(a) Select "Pictures & Videos"

(b) Take the picture and select the "Send" button

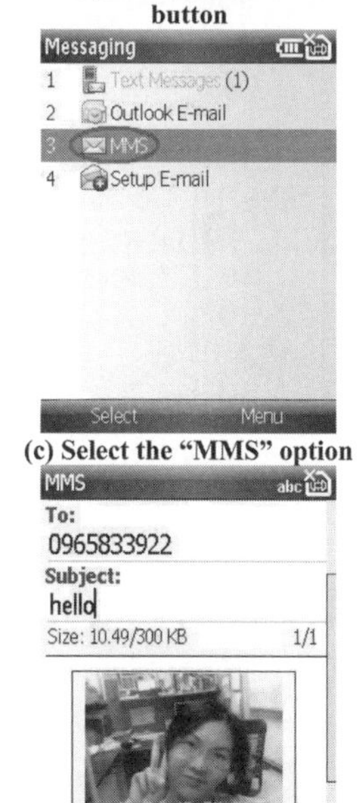

(c) Select the "MMS" option

(d) Input the phone number and subject; Press the "Send" button

Figure 2.10 MMS photo delivery (reproduced by permission of © 2007 Qisda Corporation)

2.4 IP Multimedia Core Network Subsystem

Wireless Internet has become a major trend in telecommunications. The former chair of the US FCC, Michael K. Powell stated: "I love my DSL, but I love my WiFi more. And I probably get on the Internet 40 percent to 50 percent more because of the combination of those technologies." To effectively integrate mobile technology with the Internet, UMTS introduces the IMS architecture that effectively provides multimedia services [IET02, 3GP06a]. The IMS protocols allow the telecom operators to bring attractive new services to their customers. Such protocols are namely *Session Initiation Protocol* (SIP) for signaling and Diameter for *Authentication, Authorization and Accounting* (AAA).

As illustrated in Figure 2.11, the IMS connects the GGSN (Figure 2.11(a)) with the external PDN (Figure 2.11(b)) and the PSTN (Figure 2.11(i)). To accommodate IMS, the HLR in UMTS has been evolved into the *Home Subscriber Server* (HSS; see Figure 2.11(n)). Besides the existing HLR functionalities, an HSS is equipped with Internet-based protocols for interaction with the IMS. The IMS nodes are similar to those in a SIP-based VoIP network [Col03, Sch99].

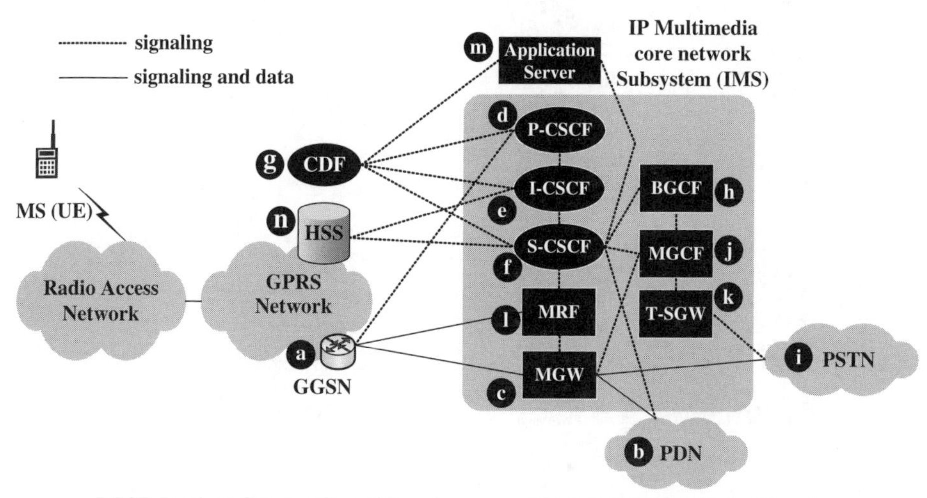

BGCF: Breakout Gateway Control Function
CDF: Charging Data Function
HSS: Home Subscriber Server
MGCF: Media Gateway Control Function
MRF: Media Resource Function
PDN: Packet Data Network
PSTN: Public Switched Telephone Network
T-SGW: Transport Signaling Gateway
CSCF: Call Session Control Function
GGSN: Gateway GPRS Support Node
I-CSCF: Interrogating CSCF
MGW: Media Gateway
MS: Mobile Station
P-CSCF: Proxy CSCF
S-CSCF: Serving CSCF
UE: User Equipment

Figure 2.11 The IMS network architecture

- The *Call Session Control Functions* (CSCFs; Figure 2.11(d), (e), and (f)) are SIP servers, which are in charge of call control and communications with the HSS regarding location information. Specifically, IMS signaling is carried out by the *Proxy CSCF* (P-CSCF; Figure 2.11(d)), the *Interrogating CSCF* (I-CSCF; Figure 2.11(e)) and the *Serving CSCF* (S-CSCF; Figure 2.11(f)) to be elaborated later.
- The *Media Gateway* (MGW; Figure 2.11(c)) transports the IMS user data traffic. The MGW provides user data transport between the UMTS core network and the PSTN (including media conversion bearer control and payload processing).
- The *Media Gateway Control Function* (MGCF; Figure 2.11(j)) controls the media channels in an MGW.
- The *Breakout Gateway Control Function* (BGCF; Figure 2.11(h)) selects the network in which the PSTN (or circuit-switched domain) breakout is to occur. If the BGCF determines that a breakout is to occur in the same network, it selects an MGCF that is responsible for interworking with the PSTN. If the breakout occurs in another IMS network, the BGCF forwards the SIP request to another BGCF or an MGCF in the selected IMS network.
- The *Transport Signaling Gateway* (T-SGW; Figure 2.11(k)) serves as the PSTN signaling termination point and provides the PSTN/legacy mobile network with IP transport level address mapping. Specifically, it maps call-related signaling between the MGCF and the PSTN.
- The *Media Resource Function* (MRF; Figure 2.11(l)) performs functions such as multiparty call, multimedia conferencing, and tone and announcement.

In Figure 2.11, an application server (Figure 2.11(m)) provides value-added IP multimedia services which reside either in the user's home IMS or in a third-party location.

When an MS attaches to the GPRS/IMS network and performs PDP context activation, a P-CSCF is assigned to the MS. The P-CSCF contains limited address translation functions to forward the requests to the I-CSCF. The I-CSCF selects an S-CSCF to serve the MS during the IMS registration procedure. This S-CSCF supports the signaling for call setup and supplementary services control. All incoming calls are routed to the destination MS through the S-CSCF. The *Charging Data Function* (CDF; Figure 2.11(g)) collects the billing and charging information from the CSCFs.

2.5 WLAN and Cellular Interworking

The 3GPP specifies WLAN and cellular interworking that extends 3G services and functionality to the WLAN access environment [3GP06c, 3GP06d]. The interworking system provides bearer services allowing a mobile user to access 3GPP PS based services through WLAN. Figure 2.12 shows the 3GPP-WLAN interworking architecture. A WLAN UE (or MS; Figure 2.12(a)) receives 3GPP PS service through the WLAN access network. The WLAN UE may be capable of WLAN access only, or

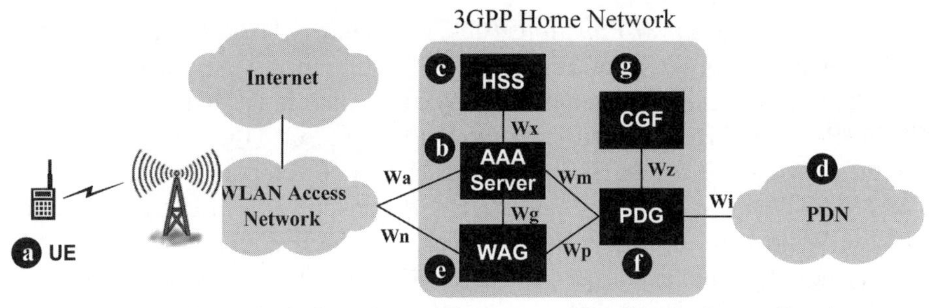

AAA: Authentication, Authorization and Accounting
HSS: Home Subscriber Server
PDN: Packet Data Network
UE: User Equipment

CGF: Charging Gateway Function
PDG: Packet Data Gateway
WAG: Wireless Access Gateway

Figure 2.12 WLAN/cellular interworking architecture

it may be capable of both WLAN and 3GPP radio access. The 3GPP AAA server (Figure 2.12(b)) retrieves authentication information and subscriber profile from the HSS (Figure 2.12(c)) through the Wx interface (implemented by the Diameter protocol). The 3GPP AAA server authenticates the WLAN UE through the Wa interface (which may be implemented by the RADIUS described in Section 5.2 or the Diameter protocol described in Section 5.3).

If the authentication is successful, the UE is authorized to access the 3GPP PS service through the *Wireless Access Gateway* (WAG; Figure 2.12(e)) and *Packet Data Gateway* (PDG; Figure 2.12(f)) via the Wn and the Wp interfaces. The WAG is a gateway that routes packets between the UE and the PDG. The WLAN UE initiates the tunnel establishment and packet encapsulation/de-capsulation with the PDG that supports WLAN access to external IP networks. Through the Wi interface, the PDG interworks with the PDN (Figure 2.12(d)). A PDN may be an external public or private packet data network or an intra-operator packet data network. Note that a WLAN UE is typically assigned two IP addresses. A local IP address is used to access the Internet directly. To access the 3GPP PS based services, the PDG also allocates a remote IP address to the UE.

When the service delivery is finished, the PDG generates the charging records that relate to user data traffic and transfers these records to the Charging Gateway Function (CGF; Figure 2.12(g)) through the Wz interface (which are also implemented by the Diameter protocol).

In Figure 2.12, the Wg interface is used by the 3GPP AAA Server to deliver routing policy enforcement information to the WAG. This information is used by WAG to securely identify the user data traffic of a particular UE and apply the required routing policy. The Wm interface enables the 3GPP AAA server to retrieve tunneling attributes and UE's IP configuration parameters via the PDG. The AAA protocol across this interface is Diameter based. The Wp interface transports tunneled user data traffic between the WAG and the PDG. A candidate protocol for the Wp interface is the GTP used in the Gn interface between the SGSN and GGSN.

2.6 Concluding Remarks

This chapter provided a brief overview of various telecommunications networks including PSTN, GSM, GPRS/UMTS, IMS and the WLAN/cellular interworking system. We described GSM-based SMS, GPRS-based WAP and MMS. Further details of the UMTS core network architecture can be found in [3GP02]. The UMTS protocol stacks are described completely in [3GP06e]. Details for GSM and GPRS can be found in [3GP05a, Lin01], and details for IMS can be found in [3GP06a, Lin05]. WLAN/cellular interworking system is specified in [3GP06c]. The complete 3GPP specifications can be found at website www.3gpp.org.

Review Questions

1. Describe the SS7 network architecture and the functions of SS7 nodes.
2. Describe the PSTN architecture. What are the differences between a central office (CO) and a tandem switch?
3. Describe the GSM architecture. What are the differences between an MSC and a gateway MSC? Which GSM nodes are connected to the billing system?
4. Describe the SMS architecture and the short message delivery procedure from an MS to another MS. Which SMS node is connected to the billing system?
5. Describe the GPRS architecture and its protocols. How many of them already exist in GSM? Which GPRS nodes are connected to the charging gateway?
6. Describe the information contained in a PDP context.
7. Describe the PDP context activation and deactivation procedures. When are the CDRs created and delivered to the charging gateway?
8. Describe the WAP architecture. What are the design guidelines for WAP? What are the disadvantages of implementing TCP/IP directly over the mobile network without WAP?
9. Describe the MMS architecture. Why is WAP utilized in this architecture?
10. Describe the functionalities of five IMS nodes. Which nodes are connected to the charging function?
11. How many kinds of CSCF are defined in IMS?
12. Describe the AAA server, the WAG and the PDG in the WLAN/cellular interworking architecture. Which nodes are connected to the charging gateway function?

References

[3GP02] 3GPP, 3rd Generation Partnership Project; Technical Specification Group Services and Systems Aspects; Architectural Requirements for Release 1999 (Release 1999), 3G TS 23.121 version 3.6.0 (2002-06), 2002.

[3GP05a] 3GPP, 3rd Generation Partnership Project; Technical Specification Group Services and Systems Aspects; Telecommunication Management; Charging Management; Charging data description for the Packet Switched (PS) domain (Release 5), 3G TS 32.215 version 5.9.0 (2005-06), 2005.

[3GP05b] 3GPP, 3rd Generation Partnership Project; Technical Specification Group Services and System Aspects; Multimedia Messaging Service (MMS); Stage 1 (Release 6), 3G TS 22.140 version 6.7.0 (2005-03), 2005.

[3GP06a] 3GPP, 3rd Generation Partnership Project; Technical Specification Group Core Network; IP Multimedia Subsystem (IMS); Stage 2 (Release 5), 3G TS 23.228 version 5.15.0 (2006-06), 2006.

[3GP06b] 3GPP, 3rd Generation Partnership Project; Technical Specification Group Core Network and Terminals; Multimedia Messaging Service (MMS); Functional description; Stage 2 (Release 6), 3G TS 23.140 version 6.14.0 (2006-09), 2006.

[3GP06c] 3GPP, 3rd Generation Partnership Project; Technical Specification Group Services and System Aspects; 3GPP system to Wireless Local Area Network (WLAN) interworking; System description (Release 6), 3G TS 23.234 version 6.10.0 (2006-09), 2006.

[3GP06d] 3GPP, 3rd Generation Partnership Project; Technical Specification Group Services and System Aspects; Telecommunication management; Charging management; Wireless Local Area Network (WLAN) charging (Release 6), 3G TS 32.252 version 6.1.0 (2006-06), 2006.

[3GP06e] 3GPP, 3rd Generation Partnership Project; Technical Specification Group Services and Systems Aspects; General Packet Radio Service (GPRS); Service description; Stage 2 (Release 5), 3G TS 23.060 version 5.13.0 (2006-12), 2006.

[3GP07] 3GPP, 3rd Generation Partnership Project; Technical Specification Group Core Network; General Packet Radio Service (GPRS); GPRS Tunneling Protocol (GTP) across the Gn and Gp Interface (Release 6), 3G TS 29.060 version 6.18.0 (2007-09), 2007.

[Che07] Chen, H.-H., *The Next Generation CDMA Technologies*. John Wiley & Sons, Ltd., Chichester, UK, 2007.

[Col03] Collins, D., *Carrier Grade Voice over IP*. McGraw-Hill, New York, 2003.

[Hol04] Holma, H. and Toskala, A., *WCDMA for UMTS: Radio Access for Third Generation Mobile Communications*. John Wiley & Sons, Ltd., Chichester, UK, 2004.

[IET02] IETF, SIP: Session Initiation Protocol. IETF RFC 3261, 2002.

[Lin95] Lin, Y.-B. and DeVries, S., PCS network signaling using SS7, *IEEE Personal Communications Magazine*, **2**(3): 44–55, 1995.

[Lin01] Lin, Y.-B. and Chlamtac, I., *Wireless and Mobile Network Architectures*. John Wiley & Sons, Ltd., Chichester, UK, 2001.

[Lin05] Lin, Y.-B. and Pang, A.-C., *Wireless and Mobile All-IP Networks*. John Wiley & Sons, Ltd., Chichester, UK, 2005.

[Sch99] Schulzrinne, H. and Rosenberg, J., The IETF Internet telephony architecture and protocols. *IEEE Network*, **13**(3): 18–23, 1999.

[Wap] http://www.wapforum.org.

3

Telecommunications Services

The American actor Tony Curtis once stated: "I can't sit around and wait for the telephone to ring." This statement indicates the significant impact of the telephone on modern everyday life. Providing efficient telecommunications services is an important issue and the main topic of this chapter. By using *Common Channel Signaling* (CCS) capabilities, the *Public Switched Telephone Network* (PSTN) can implement several popular switch functions and telephone services: one such CCS capability is *Custom Local Area Signaling Services* (CLASS) developed by Telcordia (formerly Bellcore) [Bel93]. Most of these services give customers more control, some are implemented based on the concept that people do not want to be more available. Other services are developed to assist the users in accessing their valuable calls. These services can be classified into three categories:

- *attendant functions* such as automated attendant and call conference;
- *customer functions* such as automatic call back, last number redial, call waiting, multi-connection, do-not-disturb, remainder service and status recording; and
- *system functions* such as automatic route selection, speed dialing automatic call distribution and call screening.

Some services can be implemented with more than one function type; for example, speed dialing can be implemented as either a customer function or a system function.

3.1 Automated Attendant

In traditional automated attendant service, the voice announcement answers the telephone and offers the caller a menu of choices such as "dial 0 for operator, dial 1 for sales department..." With display service interface technology, automated attendant services can be significantly enhanced; for example, the *Analog Display Services*

Charging for Mobile All-IP Telecommunications Yi-Bing Lin and Sok-Ian Sou

Interface (ADSI), originally developed in Bellcore [Jad96], provides a user with softkey access to telecom services or to internal PBX custom calling features. The ADSI offers a visual context-sensitive interface with mixed aural and visual prompts for guidance. The assistance of the display eliminates the need for remembering service code sequences and events. ADSI is an analog service because it uses analog *Frequency Shift Keying* (FSK) technology to interact with an LCD screen via low-baud rates. Bursts of modem data (specifically, FSK modulated data bursts) are sent from the service center to the user. *Dual-Tone Multi-Frequency* (DTMF) signaling is used in the reverse direction from the user to the center. Besides the standard analog telephone features, an ADSI phone has a text-based screen (20 or 40 characters wide, four to nine lines high; see Figure 3.1(a)) and a series of programmable keys (see Figure 3.1(b)) which activate service features. A server located in the PSTN offers ADSI services, downloads service scripts and controls interactive sessions. A service through an interactive session between the server and the ADSI phone can refresh and re-program softkeys in real time. For new network services implemented in the PSTN, the server downloads the service logic (in binary scripts) to the ADSI phone, which automatically executes the service logic to access the service in the PSTN.

Figure 3.1 ADSI phone: (a) text-based screen; (b) programmable keys (reproduced by permission of © 2008 Fans Telecom Inc. & Fanstel Corporation)

Telecom operators have implemented ADSI-based services for industries such as banking. They have introduced new features (including call waiting deluxe and message waiting indicator) that work exclusively with ADSI telephones and restructured ADSI services billing into value-based "packages" to stimulate customer interest.

3.2 Charging Services

Niklas Zennström of Skype promotes free telephone services: "Let me stress that Skype to Skype calls and all the features that you see today – except for

SkypeOut – will remain free." Yet besides SkypeOut, he still profits from Skype (e.g., through premium rate services), and therefore, charging always exists in telecommunications networks. Charging services provide various charging methods: *automatic alternate billing* allows a call to be billed to an account that may refer to neither the calling line nor the called line; while *split charging* allows the two parties to split the charge of the call. A well-known example of split charging occurs when called parties receive cellular phone calls in countries such as the US and China. *Free phone* allows reverse charging; that is, the costs of the phone calls are charged to the called party. 800/888 services are examples of free phone. Other charging services, such as account card calling and premium rate service, are described below.

3.2.1 Account Card Calling

Account card calling or *credit card calling* allows a subscriber to place calls to any telephone and the operator charges the costs to the account specified by the account number. The account card is either a *contact card*, which needs to be inserted into the terminal, or a *contactless card*, which is read by a wireless device. Based on the card's capabilities, account cards are classified into two types:

- A *Dumb card* only stores credit (money of some kind). The pay phone terminal contains the access routines of the card. An example of a dumb card is the prepaid telephone card illustrated in Figure 3.2(a).
- A *Smart card* has a microprocessor that performs routines to manipulate the credit information. Figure 3.2(b) illustrates a prepaid IC card that may also carry out the fraud protection procedure shown in Figure 3.3. When the card is manufactured, it is assigned a unique master key. Some card data, such as the card's serial number and the master key (Figure 3.3(a)), are the inputs to a key diversification algorithm (Figure 3.3(b)) that produces a diversified key (Figure 3.3(c)). The diversified key is fed into the pay phone terminal as the input of an authentication algorithm (Figure 3.3(d)) for fraud detection.

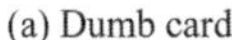
(a) Dumb card

(b) Smart card

Figure 3.2 Dumb card and smart card (reproduced by permission of © Chunghwa Telecom Co., Ltd)

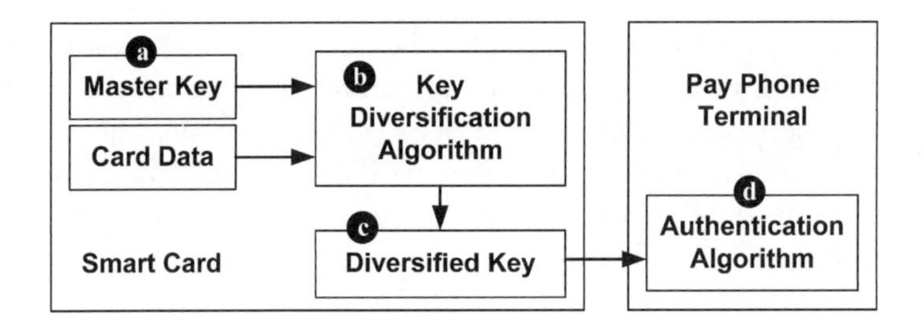

Figure 3.3 Smart card authentication procedure

3.2.2 Premium Rate Service

Premium rate service pays back part of the cost for a call to the called party, who is typically an value-added service provider. Generally, the call is an added-value call with extra charges. An example of premium rate application is the *reverse directory service* offered by the Canadian BC Tel company [Mea96]: A caller dials a toll free number and listens to a recorded voice message, which prompts the caller to enter a telephone number. The system then matches the number to information found in BC Tel's white page directory. An automated voice response system states the name and city corresponding to the entered telephone number and the caller is charged for a successful inquiry. Another example is the Internet access service that does not require a customer to input a user name and password. Instead, the customer just dials the premium rate service number and the Internet access charge is collected through the telephone bill.

Other premium rate applications include 900 number services in the US and 0204 services in Taiwan. The 900 numbers correspond to programs such as datelines, live chat lines, one-on-one talk lines, psychic lines, horoscopes, live technical support, polling and surveys, sports picks and scores, financial news and information, and stock quotes. Many new industries are profiting from the use of 900 numbers, including health care, banking and finance, and government. As an example, government agencies and organizations use 900 numbers to handle routine inquiries such as hunting and fishing licensing, background checks, passport applications, and lottery results. In Taiwan, political TV programs use 0204 numbers, which are akin to US 900 numbers, for voting as described in Section 3.7.

3.3 Routing Services

Ilka Chase, an American actress, once stated, "America's best buy is a telephone call (routed) to the right man." Routing is a fundamental task in telecommunications networks. A call is typically routed to the called party based on the number dialed; however, in enhanced services, the calling party or the switch attendant of the called party can reroute the destination of the call. This section describes several examples of routing services.

3.3.1 Automatic Call Distribution

Automatic Call Distribution (ACD) allows an incoming call to be routed to different destinations using a predefined allocation algorithm (e.g., balanced load sharing, hierarchy definitions, or fixed percentage for each destination) based on criteria such as the time of day, day of the week, location of the call's origin, calling line identity, or applicable charge rates for the destinations. Airline information services provide an example of ACD: A hypothetical airline has two information centers at New York and Hawaii, as illustrated in Figure 3.4. The working hours for the operators in both cities are from 10am to midnight and from 5am to 7pm at local times, respectively. During the New York period from 3pm to 7pm, operators in both centers are available and all customer calls are processed at the near-by centers (Figure 3.4(a)). During the New York period from 7pm to 5am, the New York center is not available and all calls are routed to the Hawaii center (Figure 3.4(b)). During the New York period from 5am to 3pm, the Hawaii center is not available and all calls are routed to the New York center (Figure 3.4(c)). This service is typically implemented by using the time-of-day feature in the 800/888 approaches.

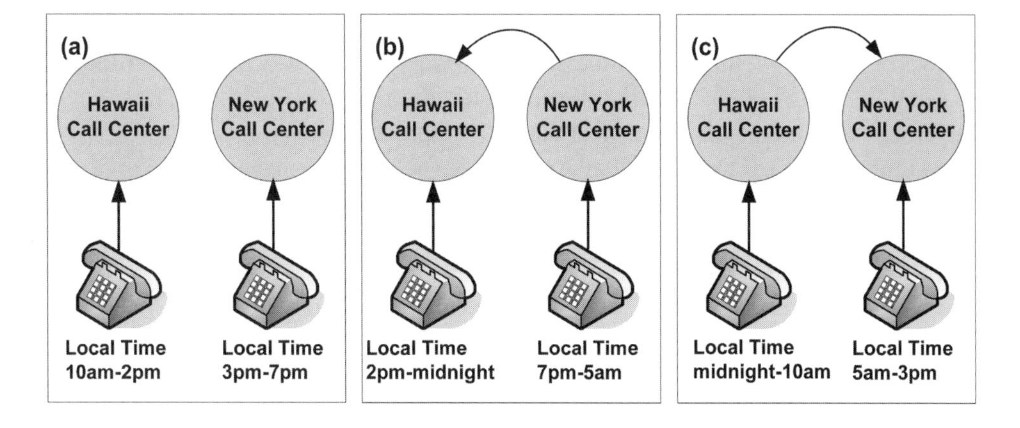

Figure 3.4 Call routing between two call centers

In *uniform call distribution*, an incoming call is routed to the first available agent in the call center. If no agent is available, the call is routed to an appropriate announcement and/or held in a queue until an agent is available. The queued calls are typically answered on a first-come, first-served basis. The ACD may also advise a queued customer when the next agent will be available [Hsu06]. In addition to all queuing and routing functions, the call distribution mechanism provides statistical reporting tools. For example, the reports can be used to determine the appropriate number of staff required for specific weeks and months in the future.

3.3.2 User-defined Routing

User-defined routing allows a subscriber to define the route of outgoing calls through private, public, or virtual facilities. Routing criteria include least cost routing, route quality, best pattern match, circuit routing, domain priority, answer seizure ratio, and load balancing. Routing criteria and prioritization are configurable for each ingress point into the PSTN. *Automatic route selection* or *least cost routing* delivers outgoing calls using the least expensive route. Specifically, the calls are connected to available carriers at the lowest available cost (by considering the termination costs, time-of-day availability and so on). User-defined routing is typically supported through a call routing server for *Voice over IP* (VoIP). In IMS, the breakout gateway control function (Figure 2.11(h)) is responsible for selecting the routing from IMS to other telecommunications networks.

Call forwarding transfers an incoming call to another telephone number. Various events can trigger this service: If a line is busy, *call forwarding on busy* routes an incoming call to another telephone number. If the calling number is on a pre-selected list, *selective call forwarding on busy* forwards the call. If the called party does not reply, *call forwarding on no reply* forwards a call to another number. *Follow-me diversion* or *unconditional call forwarding* enables the user to dynamically control the call forwarding mechanism from a remotely located telephone. A customer, for instance, can initiate the mechanism while she is away from her home, and all calls will follow her while she is traveling. When the call forwarding service is activated, the forwarded line will ring once, to remind the customer that the call is being redirected. The customer should remember to cancel the unconditional call forwarding feature when she returns to her original telephone; otherwise, she will no longer receive any calls.

Charging for call forwarding can be subtle. For example, if customer A forwards her phone number to a mobile number, then a caller placing a call to customer A may incur higher rates due to the subsequent use of the mobile telephone network. On the other hand, customer A usually incurs all related charges, including long distance.

3.4 Dialing Services

In a typical call setup procedure, the user dials the destination number to connect the call. Several enhanced services offer assistance to speed the dialing process; for instance, automatic dialing, automatic callback, last number redial, direct inward dialing and speed dialing.

3.4.1 Automatic Dialing

Automatic dialing automatically dials a pre-programmed phone number. *Hotline* services, for instance, automatically route calls to given destinations. A hotline service engages if, after picking up the handset, the calling party does not dial the first digit within the hotline timeout period. Another example of automated dialing is an "auto-dialer" that automatically calls out to a list of phone numbers, leaves personalized messages on answering machines, and plays voice messages to the called parties. When the called party plays the voice message, she can press a special key on her keypad and be transferred directly to one of the agents of the autodialer. An example of the user profile for an incoming customer that will be transferred among the agents is illustrated in Figure 3.5.

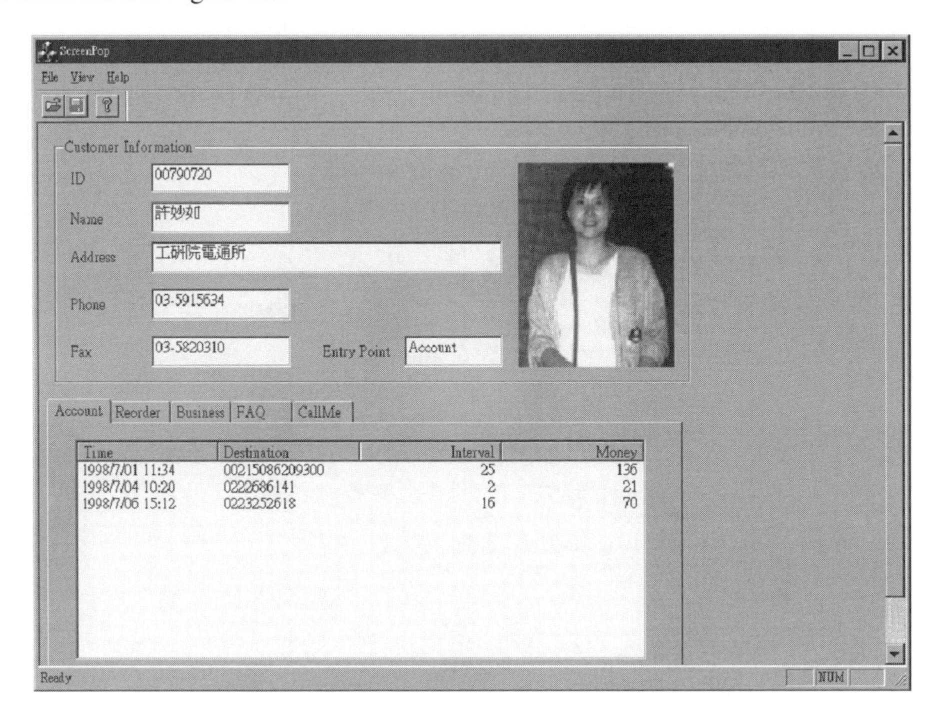

Figure 3.5 An example of user profile transferred among the agents

3.4.2 Speed Dialing

Speed dialing allows a user to access frequently used numbers by dialing abbreviated codes. Speed dialing operates even if different *Service Switching Points* (SSPs) serve the calling and the called party lines. Speed dialing is either *individual abbreviated* or *system abbreviated*.

- For individual abbreviated speed dialing, a user programs a list of abbreviated codes into her telephone. The subsequent procedure is illustrated in Figure 3.6(a): The calling party A dials the number "12", which translates into the destination number "1234567" at the telephone set. This number is then sent to the SSP, which routes the call based on the received number. The same abbreviated code is typically translated to different telephone numbers for different telephone sets: when another calling party B dials the same number "12", it may translate to number "7654321" as shown in Figure 3.6(a).

- For system abbreviated speed dialing, a given customer group has access to a list of abbreviated dialing locations. In Figure 3.6(b), when the caller dials the number "12", this abbreviated code is sent to the SSP. After the SSP maps and locates the address, the destination address "1234567" is found, and it routes the call to the intended party. In this approach, all calling parties dialing the number "12" are connected to the same number "1234567". If the SSP also checks the caller ID, then the same abbreviated number can be mapped to different called numbers as in the individual abbreviated approach.

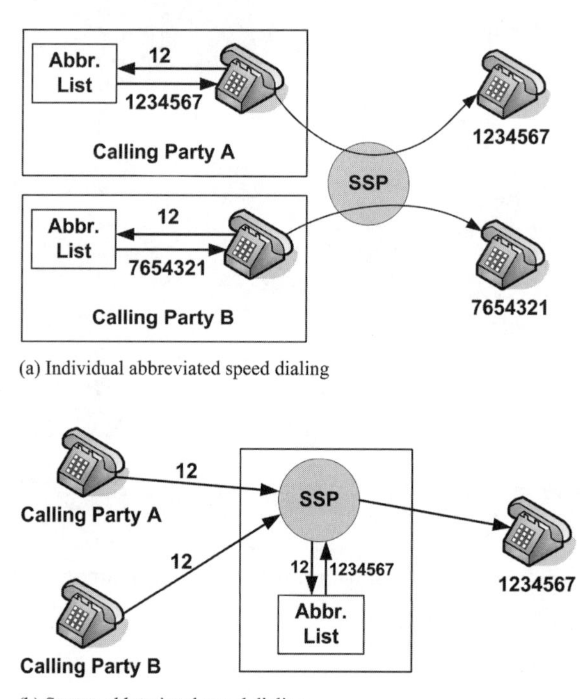

(a) Individual abbreviated speed dialing

(b) System abbreviated speed dialing

Figure 3.6 Individual and system abbreviations

3.4.3 Callback, Redial and Inward Dialing

Automatic callback or *automatic ring back* allows the caller to request a callback, while *automatic callback on busy* requests a callback when the called party is busy. After the called party terminates the previous call (goes on-hook), the caller telephone is signaled and can go off-hook to automatically initiate another call to the called party again. *Automatic callback on absent* allows the caller to request a callback if the called party does not reply. After automatic callback on absent is enabled, if the called party finishes any call activity (either originating a call or receiving a call), a call is automatically set up between the calling and the called parties.

Last number redial or *automatic recall* automatically redials the last dialed number. This service has two options: In the first option, if the called party does not reply to a call, the caller dials a short code or presses a function key to repeat the last dialed number. Once the call connects, the service no longer remembers the last dialed number. In the second option, the service always remembers the last dialed call whether it connected or not.

Direct Inward Dialing (DID) gives a PBX extension user a direct telephone number besides the extension number. With DID, a telecom operator allows a corporation to subscribe to fewer leased lines than extensions, where every extension is assigned a unique telephone number (a DID number) that can be reached from outside the corporation. The external leased lines are switched to the internal extensions through a PBX. When a DID call from the PSTN arrives, the PBX maps the DID number to the corresponding extension.

3.5 Screening Services

Paul Auster begins his novel *City of Glass* with the famous first lines: "It was a wrong number that started it, the telephone ringing three times in the dead of night, and the voice on the other end asking for someone he was not." In a typical call setup procedure, the called party is alerted when the calls arrive. To avoid bothersome or suspicious calls, *screening services* allow a call party (either the calling party or the called party) or the telephone owner to block access between specific phone lines.

Call blocking (*call screening* or *restricted calling*) allows a customer to restrict certain types of calls such as those from 900 entertainment services in the US. The customer may also specify restrictions on outgoing calls by creating a screening list and, optionally, by entering time-of-day specifications. For example, after hanging up with a bothersome telemarketer, the customer can block calls from that number by pressing a simple code on the dial pad. *Security screening*, another blocking function, performs a security procedure in the network before a caller gains access to the subscriber's network, systems or applications. This service is particularly useful when it is integrated with credit card calling, where each erroneous attempt to enter the PIN number is recorded and ultimately used for screening purposes. An example of security screening is *forced account code* that asks the caller to enter a verified account code before making an outgoing call.

Both *receive-only* and *do-not-disturb* screening mechanisms prevent access to a called party. The receive-only screening mechanism is associated with the calling line: When a user activates the mechanism, the telephone line can only receive incoming calls. On the other hand, when a called party activates the do-not-disturb service, all callers automatically receive a busy signal. *Do-not-disturb with PIN* service forwards all incoming calls to a recorded message from the telecom operator which informs the caller that the called party does not wish to be disturbed at this time. Callers (with touch-tone telephones only) with the correct *Personal Identification Number* (PIN) can override the service and proceed with their calls; therefore, the called party can still receive important calls from selected friends or relatives. In the *do-not-disturb telemarketing* service, all incoming calls are forwarded to a telecom recording, such as "You have called a number which does not accept calls from telemarketers. All other callers may press '1' if they wish to complete the call." Since most telemarketers use autodialers, they are unable to respond to the do-not-disturb service's request for a digit '1' input. In the US, a telemarketer who completes the call would be in violation of the *Telephone Consumer Protection Act* (TCPA) and may be punishable by up to US$1500 per incident. The do-not-disturb service takes precedence over call forwarding and/or call waiting features.

3.6 Interrupt Services

The interrupt services allow the call parties or an attendant to redirect or control the call connection during a conversation. *Private call* prevents all interruptions and intrusions from either the attendant or other telephones. *Call holding* allows a call party to place the other party on hold and either deal with the incoming call or make a new call, and then switch from one call to the other without terminating either of them. *Call transferring* allows a call party to transfer the call to another telephone. To transfer a call, the call party presses the flash button or switch hook and dials a special code (e.g., #90). Then the call party dials the phone number of the intended recipient of the call before dialing another special code (e.g., #). When the call party hangs up the phone, the transfer will take place.

Executive override allows specified users to break into an established call based on its precedence (i.e., the priority level associated with the call). Precedence assignment represents an ad hoc action in that the user chooses whether or not to apply a precedence level to a call attempt. Precedence priorities include *executive override*, *flash override*, and so on. When the executive override (the highest) precedence level preempts a lower precedence call, the executive override call changes its precedence level to flash override (next highest level), so a subsequent executive override call can preempt the first precedence call.

Call waiting allows a customer to answer a second call when the customer is already engaged with a call. The steps of line re-direction in the call waiting service are illustrated in Figure 3.7 and explained as follows:

Step 1. Initially, user A is engaged in a call with user B.

Step 2. User C attempts to call user A. Since user A has subscribed to the call waiting service, the SSP forwards the call waiting DTMF signal to user A.

Step 3. User A replies to the SSP with a connection switching signal.
Step 4. The SSP switches the line to connect users A and C.

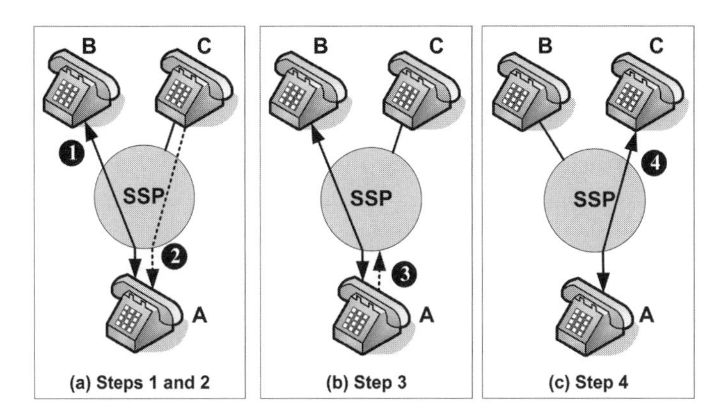

Figure 3.7 The call waiting procedure

In Step 3, the connection switching signal from user A is a *flash-hook* signal that is typically implemented with the *switch hook flashing* technique, whereby the user generates the flash-hook signal by momentarily depressing the switch hook. This operation is time-sensitive: If the depressed period is too short, the signal is not recognized; if the depressed period is too long, the call may be disconnected.

3.7 Mass Call

Mass call allows a telecom operator to temporarily allocate a directory number to a user. Each time a caller dials the allocated number, an announcement asks the caller to input more digits to indicate her preference from a menu of choices. This procedure is similar to the automatic attendant service explained above; however, mass call is typically used to handle events involving a large number of simultaneous calls, such as voting during a television show. In *televoting*, a caller either dials a specific number according to her choice, or dials a number and uses the automated attendant menu service to indicate her choice.

Consider the following televoting procedure: Two candidates, John and Jenny, are assigned the (voting) telephone numbers of 1111111 and 2222222, respectively. If a voter selects Jenny, the voter initiates the following steps (see Figure 3.8):

Step 1. The voter dials the number 2222222. This call request is received by the SSP.
Step 2. The SSP sends an SS7 signaling message to the *Service Control Point* (SCP). The message indicates a vote for Jenny.

Step 3. The SCP accesses the voting database and advances Jenny's counter by one increment.

Step 4. The SCP reports the result to the SSP.

Step 5. The SSP may deliver a voice announcement to the voter to indicate a successful vote.

Step 6. The SCP may periodically report the televoting statistics to the *Service Management System* through a data network (e.g., X.25 or Internet). The results are then broadcast on a real-time TV program.

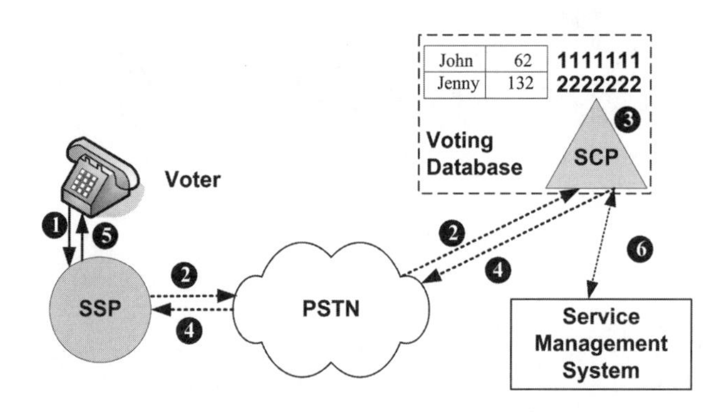

Figure 3.8 Televoting procedure

The architecture illustrated in Figure 3.8 efficiently supports the mass call service. Unlike normal phone calls, televoting does not consume any voice trunks because the voting procedure is carried out by packet-switched oriented signaling messages through the SS7 network (see Steps 2 and 4 of Figure 3.8).

3.8 Universal Personal Telecommunications Number

A *Universal Personal Telecommunications* (UPT) number supports *personal mobility* by tracking the locations of the UPT users. Personal mobility allows a UPT user to access telecommunications services with any terminals (e.g., telephones) in any locations within the service area. The UPT provider assigns each user an internationally unique UPT number, which is an ITU-T recommendation E.164 number [CCI91]. The provider receives connections from the international telephone network and forwards each called number to the UPT user's real number. When accessing telecommunications services, the UPT user may be limited by network restrictions, terminal capabilities, or regulatory requirements. An example of a UPT provider is the international association VISIONng, which has been allocated a UPT number range within the +878 country code in the US.

UPT user service profile and *UPT terminal profile* are introduced to associate a UPT user with any terminal (telephone). The UPT service profile contains information such as subscriptions to basic and supplementary services and call-routing preferences. Each UPT service profile is associated with a single UPT number. As shown in Figure 3.9, the UPT service profile collocates with the SCP. A record in the service profile indicates the location (the SSP) of the corresponding UPT user. A record in the terminal profile indicates the relationship between a terminal connected to the SSP and the UPT user concurrently using that terminal.

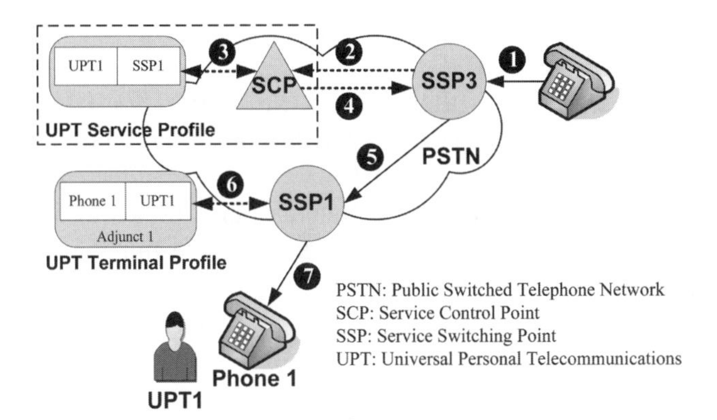

(a) UPT call delivery

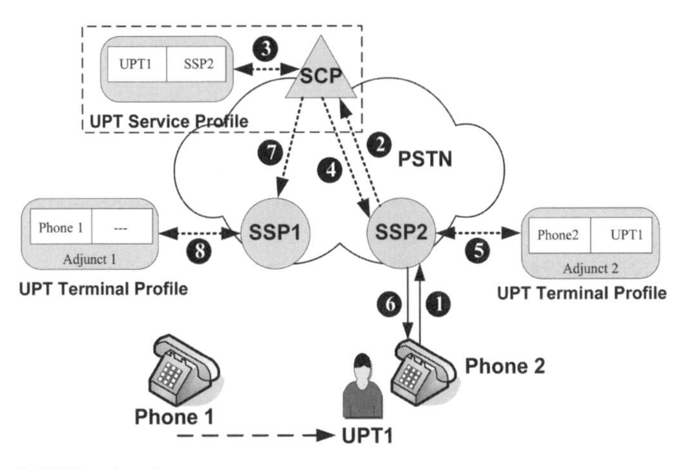

(b) UPT registration

Figure 3.9 The UPT location update and call delivery

The service profile resides in an *adjunct* nearby the corresponding SSP. The adjunct is similar to an SCP, except that the adjunct processor has small capacity and uses

TCP/IP (through high-speed Ethernet) instead of SS7. The processing speed of the adjunct processor is generally faster than the SCP; thus, the adjunct processor is appropriate for small service providers (e.g., those with one SSP that do not need SS7 connections).

The UPT call delivery procedure is described in the following steps (see Figure 3.9(a)):

Step 1. The calling party dials the UPT number UPT1. This call request is sent to the switch SSP3.

Step 2. SSP3 recognizes the UPT number and launches an SS7 query to the corresponding SCP.

Steps 3 and 4. The SCP identifies the location (i.e., SSP1) for UPT1 in the UPT service profile and sends the location information to SSP3.

Step 5. SSP3 sets up a trunk to SSP1.

Step 6. SSP1 recognizes that the call setup corresponds to a UPT user and queries the terminal profile to identify the terminal (i.e., Phone 1) for UPT1.

Step 7. SSP1 connects the call to Phone 1.

In Figure 3.9(a), the dashed lines (2, 3, 4, and 6) are signaling paths and the solid lines (1, 5, 7) are voice paths. When the UPT user moves to a new location, she should register with a new terminal at that location so that the UPT network can track her location. Figure 3.9(b) illustrates the following *registration* procedure:

Step 1. When the user with number UPT1 moves from SSP1 to SSP2, she changes the terminal from Phone 1 to Phone 2. The user initiates the registration procedure by sending a request to SSP2 (typically through Phone 2).

Step 2. SSP2 forwards this registration request to the SCP.

Step 3. After an authentication procedure, the SCP updates the UPT1's record in the service profile (changing the location from SSP1 to SSP2).

Step 4. The SCP sends an acknowledgment message to SSP2.

Steps 5 and 6. SSP2 modifies the terminal profile by associating Phone 2 with UPT1 and forwards the successful registration acknowledgment to UPT1.

Steps 7 and 8. Concurrent with Step 4, the SCP sends a deregistration message to SSP1. SSP1 modifies the terminal profile by removing UPT1 from Phone 1's record.

With the above registration and call delivery procedures, a UPT user may access telecommunications services with a unique UPT number.

3.9 Interactive Voice Response Techniques

The services described in the previous sections can be significantly enhanced with the *Interactive Voice Response* (IVR) techniques [Pec95]. User responses in most existing services rely on DTMF signaling, where users press the telephone buttons to decide their selections. DTMF signaling is tedious because of the limited vocabulary (0–9, *, #) available to the user. A better alternative is IVR, which is based on the automatic speech recognition technology.

IVR systems are typically used in bank balances, order placement, and ticket booking. In 1989, Bell Northern Research first applied IVR to the automated alternative billing service; at the time, the system recognized only a small vocabulary (such as "yes", "no", and some synonyms). In 1992, AT&T introduced an IVR-based call service system: The system uses *word-spotting* (recognition of key words in the midst of additional acoustic material) and *barge-in* (talkover or echo cancellation) techniques. Most IVR technologies are based on the *Hidden Markov Modeling* (HMM), which is capable of capturing the variations in pronunciation of the words. An HMM database stores a large number of speech patterns from many speakers to facilitate speech recognition from a diverse clientele. Modern IVR systems use natural language speech recognition to interpret callers' questions; some also rely on human agents to aid the speech recognition mechanism.

A customer (typically a corporation) can either purchase an IVR platform to use at a specified location or pay a monthly charge for an IVR supplied by a telecom operator. IVR systems are often criticized as being difficult to use and unresponsive to callers' needs. Therefore, it is essential that a properly designed IVR system should connect callers to their desired services promptly.

3.10 Other Telephone Services

This section introduces some popular services not described in the previous sections. *Voice call*, for example, allows a user to receive calls hands-free. With voice call enabled, calls are connected immediately, following a single ring tone heard by the called party. With voice call disabled, the telephone rings to indicate an incoming call and is manually picked by the called party to answer the call.

Multi-connection services relax the one-to-one call connection restriction. For example, the *paging* service allows a caller to broadcast a page to all assigned telephone members of the selected paging group.

Conference call or *party line* allows multiple users to connect simultaneously. The calling party may specify the time of the meeting and the meeting's members; the system then automatically connects the members at the specified time. The calling party may add more members during the conference call. In *dial-in teleconferencing*, a subscriber books the meeting in advance (through a telecom operator website) and specifies a conference name, the number of participants, date and time, and the duration of the conference. Conference calls are most commonly used in business, as they are ideal for large group conferences such as sales meetings and staff meetings, workshops, or brainstorming sessions. Another conference call application, *party line*, allows customers to dial a specified telephone number to talk to other party-line participants, and perhaps meet new people. Traditionally, conference call services are supported by hardware (a conference bridge) installed in the SSP, and a telecom operator charges setup fee. However, many modern service providers (such as Skype) offer conference calls on the Internet for free.

Calling number services, such as *caller ID* and *calling number delivery*, display the telephone number of the caller; *calling name delivery* displays both the caller's name and her ID. The called party's telephone answering equipment may display the caller's identity in various ways: it may appear on a telephone or computer screen, be stored automatically in a database, or be recorded as an audio message. Figure 3.10 shows caller ID product that displays the caller ID for call waiting and voice mail services. Many services described in the previous sections incorporate the caller ID feature. The calling number services enhance privacy for the called party, reduce obscene or harassing calls, and improve the response time of police and firefighters. On the other hand, the services discourage calls to anonymous help hotlines, expose

Figure 3.10 A caller ID product that displays the caller ID for call waiting and voice mail services (reproduced by permission of © 2008 Fans Telecom Inc. & Fanstel Corporation)

unlisted numbers, and erode the "expectation of anonymity" of the calling public. The identity disclosure associated with caller ID sparked a legal concern for American legislators: In Pennsylvania, a judge suspended Bell Atlantic's calling number services, saying that it may violate state wiretap laws. With reverse directories, any interested party can readily infer a caller's identity from a given phone number. Businesses, for example, can collect and use callers' numbers to build profiles about existing and prospective customers. *Calling number delivery blocking* ensures that the caller's number is not disclosed to any called party. Blocking technologies can be implemented on either per-call or per-line basis. A telecom operator may override the block when the calling party dials to an *Internet Service Provider* (ISP) to log on to the Internet.

Voice mail records messages when a called party is away from the phone or engaged in another call. An advanced mailbox system supports one mailbox for all voice, fax, and email messages. The system can answer many phones at the same time, send and forward messages to multiple voice mailboxes, add a voice introduction to a forwarded message, store voice messages for future delivery, and notify a user of new messages through a telephone or paging service. The system can also personalize services: It can store incoming voice messages in mailboxes associated with each user's phone number, transfer specified caller to another phone number for personal assistance, and play different message greetings to different callers. Voice mail systems may be integrated with automated attendant facilities, which would enable users to place calls to a main business number to access the business's directory or self-route the calls to a specific department, an extension number, or an informational recording in a voice mailbox.

In 1996, Vietnam introduced the so-called *virtual phone* service using voice mail technology [Yab96]. Virtual phone service does not require dedicated phone lines, which were too expensive to be widely installed in the country at the time of implementation. Instead, each subscriber is assigned a virtual phone number: callers leave voice messages in a mailbox associated with the called number. Through a password identification process, the owner of the voice mailbox can periodically check her mailbox from any phone anywhere in the world. A modern voice mail system is capable of *presence management* to detect Internet connectivity and the availability status of the called party to exchange real-time messages; it also includes *buddy list* directories to allow only authorized calling parties to initiate real-time text messaging exchanges with the called party.

Still further services include *status recording*, in which customers or attendants can access call information, and *malicious call identification*, in which subscribers can record calls that are of a malicious nature. The record includes the time of the call and the identifications of the call parties. *Station message detail recording* provides a real-time, detailed call record for every PSTN call processed by the system, while *customer originated trace*, when activated, automatically traces the last incoming call. Police often use this feature to investigate threatening or obscene calls. Through a dedicated data link, *bulk calling line identification* allows the PBX customer to receive call-related information for an incoming call outside the PBX, such as caller's number, the called party number, or the time and date of a call.

Reminder service allows a user (or attendant) to program a reminder call at a specified time; when the time arrives, the telephone rings. Hotels use this technology for *wake-up call*. *Call pickup*, yet another telephone service, allows a user to answer a call that is ringing at another telephone, while *distinctive ringing* distinguishes calls from pre-identified numbers using a special alerting signal.

3.11 Mobile Telecommunications Services

The former U.S. congressman and retired football player Steve Largent said: "I think we'll quit calling *mobile device* a phone since it'll have so many more functions to it." Today, UMTS, WiMAX and other broadband wireless systems provide bandwidths that are sufficient for most Internet applications such as web browsing, image transfer, and content delivery for video clips and MP3 music files [Say02]. To characterize mobile data applications, the 3GPP defines four QoS types: conversational, streaming, interactive, and background [3GP06, Ols06]. The characteristics and applications of each type are listed in Table 3.1.

Table 3.1 QoS types in UMTS

Types	Characteristics	Applications
Conversational	• Delay and jitter controlled • Constant bit rate • Some bit errors allowed	Voice, video telephony, videoconferencing
Streaming	• Jitter controlled • Near constant bit rate • Some bit errors allowed	Video, audio
Interactive	• Enables question/answer exchange • Low or no tolerance of errors • Variable bit rate	Web browsing, interactive email, interactive game
Background	• No time constraint • Low or no tolerance of errors • Variable bit rate	FTP, email downloading, messaging

Based on the above description, charging for mobile services can be categorized into two classes: event-based charging and session-based charging. Event-based charging implies that a chargeable event is defined as a single transaction, e.g., sending of a multimedia message, or downloading a ring-tone from the website. This chargeable event is mapped to a single *Charging Data Record* (CDR). In contrast, session-based charging is used for a service session, such as a GPRS *Packet Data Protocol* (PDP) session or an IMS session. A service session generates multiple chargeable events and creates one or more CDRs. In general, session-based services are charged by the served time or packet volume transferred during the session. Table 3.2 lists some applications that are typically charged by event, time, or traffic volume [Nok04].

Table 3.2 Charging methods of mobile services

Mobile services	Event	Time	Traffic volume
Real-time services (voice, video, streaming)		x	
Non-real-time services (messaging, gaming)	x		
Content downloading (ring tone, java game, movie clip)	x		
Corporate intranet access		x	x
Internet access (browsing, email, streaming)			x

Recently, mobile device manufacturers have developed multimedia-ready mobile phones with enhanced performance and features. These devices provide user-friendly interfaces, support new multimedia content formats, and enable access to new mobile multimedia services, described as follows [3gc]:

- *Video telephony*: A broadband wireless network offering high data rates to improve video quality. Using a mobile phone with a video camera, a mobile user can place video phone calls through the *Third Generation* (3G) mobile network. Video telephony, which traditional PSTN system does not effectively support, is often considered as a 3G killer application. Figure 3.11 illustrates a video phone call from a 3G handset to a desktop PC.
- *Ring-back tone and video services*: Ring-back tone is an audio sound heard by a calling party while she waits for the connection to a called party to be completed. Traditional ring-back tones are simple beeping sounds. Today, personalized ring-back tone services allow a subscriber to easily configure the personalized ring-back tones from a library of offered songs through a web interface or by pressing special keys on the phone keypad. The ring-back tone concept has been extended to ring-back video that can be shown on the called MS's screen to alert the called party. Figure 3.12 shows an example of personalized ring-back tone configuration webpage.

Figure 3.11 A Video phone call from (1) a 3G mobile handset to (2) a desktop PC

(1) **Search the RBT by keyword**

(2) **RBT code**

(3) **RBT name**

(4) **Singer**

(5) **Provider**

(6) **Price**

(7) **Free trial**

(8) **Download**

鈴聲下載

一般鈴聲 關鍵字搜尋

鈴聲類別：西洋音樂 按 最新鈴聲 排序 至第 1 頁 (1)

手機直撥900, 按4後輸入代碼即可快速設定您的900來電答鈴喔!

西洋音樂

(2) 代碼	(3) 鈴聲名稱	(4) 歌手名稱	(5) 提供廠商	(6) 價位	(7) 試聽	(8) 下載	點歌送禮	原音鈴聲	全曲
324860	Calendar Girl	Stars 繁星	e7play美答鈴	15元					
324859	Soft Revolution	Stars 繁星	e7play美答鈴	15元					
324858	Celebration Guns	Stars 繁星	e7play美答鈴	15元					
324857	He Lied About Death	Stars 繁星	e7play美答鈴	15元					
324856	The First Five Times	Stars 繁星	e7play美答鈴	15元					
324855	Sleep Tonight	Stars 繁星	e7play美答鈴	15元					
324854	One More Night	Stars 繁星	e7play美答鈴	15元					
324853	What I'm Trying to Say	Stars 繁星	e7play美答鈴	15元					
324852	The Big Fight	Stars 繁星	e7play美答鈴	15元					
324851	Reunion	Stars 繁星	e7play美答鈴	15元					

頁次:1/1917

|1|2|3|4|5| 下五頁

Figure 3.12 A personalized ring-back tone (RBT) configuration webpage (reproduced by permission of Far EasTone Telecommunications Co., Ltd)

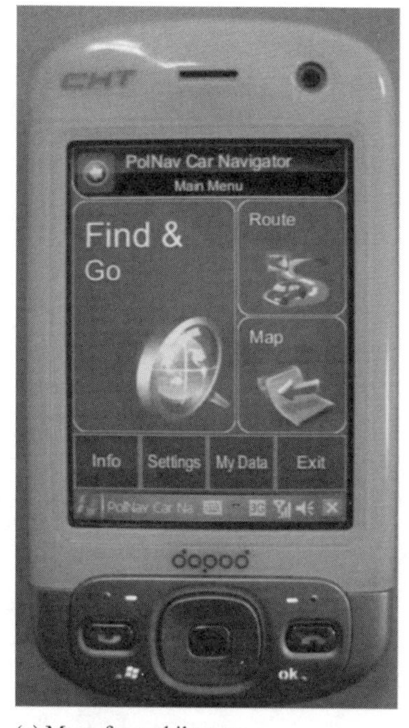

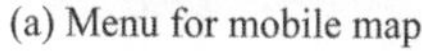
(a) Menu for mobile map

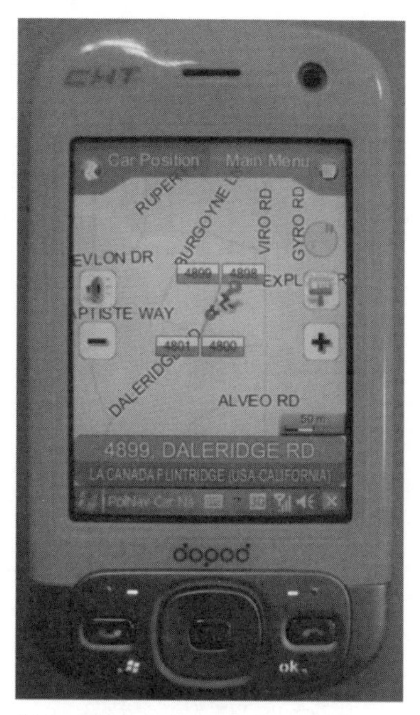

(b) Real-time road information

Figure 3.13 GPS information downloading (reproduced by permission of Polstar Technologies Inc.)

- *GPS information downloading*: In GPS navigation applications, users can download maps and live traffic information from a central server to a mobile device. Figure 3.13 illustrates a GPS information application example.
- *Instant messenger*: Mobile users can access instant messaging for communicating with friends and families. Windows Live Messenger (MSN Messenger) is the world's most popular instant messaging service. Customers have been able to enjoy the familiar look and feel of MSN to chat on their mobile devices.
- *Interactive mobile games*: Traditionally, mobile users download games to their mobile devices. With a high-bandwidth mobile telecommunications network, they can now play interactive mobile games with remote peers. These games allow a player to see an opponent's movements in real time. Figure 3.14 illustrates the Connect6 chess game [Wu05] implemented in Nokia S60 platform. Figure 3.14(a) shows how to invite an opponent to play the game; Figure 3.14(b) shows how the opponent accepts the game; Figure 3.14(c) and (d) show the snapshots of the game in progress.

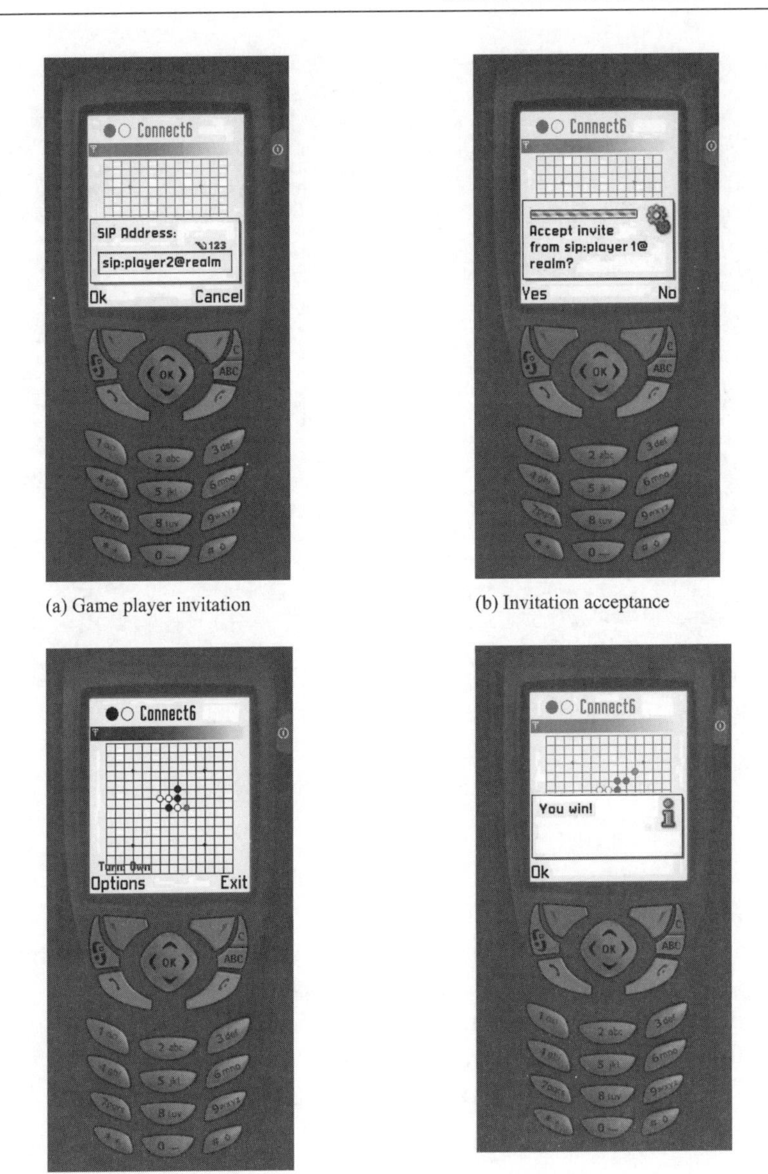

(a) Game player invitation

(b) Invitation acceptance

(c) Game playing

(d) Game termination

Figure 3.14 Connect6 mobile game developed in Nokia S60 platform

3.12 Concluding Remarks

In this chapter, we described a number of telecommunications services: First, we described traditional PSTN services, including routing services, dialing services, screening services, interrupt services, and mass call; then we elaborated on some

3G multimedia services including video telephony, GPS navigation application, and mobile games. For more details about 3G applications, readers are referred to the web site [3gc]. We emphasize that integrating wireless and mobile technologies with IP core networks will enable service providers to offer many advanced IP-based multimedia services to mobile subscribers. Integrating these technologies requires substantial enhancements to the functionality of both mobile terminals and the complicated human–machine interface (HMI).

Bjarne Stroustrup, the inventor of the C++ language, once said, "I have always wished that my computer would be as easy to use as my telephone. My wish has come true. I no longer know how to use my telephone." It is important that when new data services are introduced, the MMI must be designed to maintain, as much as possible, the simplicity of "telephone characteristics". This issue is still open for further study. Clearly, the touch screen concept of Apple iPhone will be a main direction for research.

Review Questions

1. Define attendant, customer, and system functions for telecommunications services.
2. How can VoIP applications such as Skype support ADSI service? What kinds of mobile handsets can conveniently implement ADSI service? (Hint: How do you operate an iPhone?)
3. Describe the differences between automatic alternating billing and split billing.
4. Describe the differences between dumb card and smart card.
5. Describe the smart card authentication procedure.
6. Give an example for premium rate service.
7. Describe a call center architecture that allows the call agents to work at home.
8. Describe the options for call forwarding.
9. Give examples for automatic dialing.
10. Layout a web-based user interface for business conference call. Then modify this interface for the party line service.
11. How do you implement voice mail by using IVR?
12. Give a scenario to illustrate how buddy list can be used in voice mail.
13. Which speed dialing service(s) will involve telecommunications billing? Individual abbreviation, system abbreviation, or both?
14. Describe the difference(s) between automatic callback on busy and automatic callback on absent.
15. Describe the difference between call forwarding and UPT service.

16. Voice mail feature provided by telephone companies is more popular for cellular phone service than for fixed network service. Why?
17. Which services described in this chapter cannot be charged by a telephone company? For example, a customer can purchase a caller ID product (see Figure 3.10) that performs all calling number-related services. In this case, how can the telephone company charges the customer (if it wants to)?
18. For services described in this chapter, which of them can be charged by per transaction? Which of them can be charged by flat rate?
19. Describe two 3G applications that are not introduced in this chapter.

References

[3gc] http://www.3g.co.uk/3GForum/.

[3GP06] 3GPP, 3rd Generation Partnership Project; Technical Specification Group Service and System Aspects; Quality of Service (QoS) concept and architecture (Release 6), 3G TS 23.107 version 6.4.0 (2006-03), 2006.

[Bel93] Bellcore, *A Guide to New Technologies and Services*, Technical Report SR-BDS-000828, Bellcore, 1993.

[CCI91] CCITT, *Numbering plan for the ISDN era*, Technical Report Recommendation E.164 (COM II-45-E), ITU-T, 1991.

[Hsu06] Hsu M.-T., Lin, Y.-B., Li B., and Chang M.-F., A SIP-based call center with waiting time prediction, *Journal of Internet Technology*, **7**(4): 313–322, 2006.

[Jad96] Jadoul, M., Interactive services on the public network, *Telecom Asia*, **7**(11): 24–29, 1996.

[Mea96] Meade, P., Canadian Telco officers use a handy reverse directory, *American's Network*, p. 32, May 15, 1996.

[Nok04] Nokia, *Service Charging in Intelligent Edge,* 2004.

[Ols06] Olsen, B.T. *et al.*, Technoeconomic evaluation of the major telecommunication investment options for European players, *IEEE Network*, **20**(4): 6–15, 2006.

[Pec95] Peckham, J., IVR and the art of conversation. *Telecommunications*, **4**: 153–159, September 1995.

[Say02] El-Sayed M. and Jaffe, J., A view of telecommunications network evolution, *IEEE Communications*, **40**(12): 74–81, 2002.

[Wu05] Wu, I.-C., Huang, D.-Y., and Chang, H.-C., Connect 6, *ICGA Journal*, **28**(4): 234–241, December 2005.

[Yab96] Yablonski, D., Voice messaging in Vietnam: virtual for a real challenge, *Telecommunications*, pps. 98–102, 1996.

4

GPRS Tunneling Protocol Extension

When a *Mobile Station* (MS) accesses the data services through the *General Packet Radio Service* (GPRS), the *Charging Data Records* (CDRs) are generated based on the charging characteristics (data volume limit, duration limit and so on) for that service. As illustrated in Figure 2.5, the *GPRS Support Nodes* (GSNs) and the *Charging Gateway* (CG) are connected through the Ga interface. In a GPRS session, the CDRs are generated by the GSNs, and are sent to the CG via the *GPRS Tunneling Protocol Extension* (GTP'). A GSN only sends the CDRs to the CG(s) in the same UMTS network. A CG analyzes and possibly consolidates the CDRs from various GSNs, and passes the consolidated data to a billing system. This chapter describes how the GTP' protocol is utilized between the GSNs and the CG.

4.1 The GTP' Protocol

The GTP' protocol is used for communications between a GSN and a CG, which can be implemented based on UDP/IP or TCP/IP. In some commercial products, the GPRS default configuration specifies UDP, which is a connectionless-based protocol that is considered an unreliable transport method but can yield better performance. On the other hand, TCP is a connection-based protocol that provides reliable transmission through packet acknowledgment. GTP' utilizes some aspects of the *GPRS Tunneling Protocol* (GTP) defined in 3GPP TS 29.060 [3GP07]. Specifically, *GTP Control Plane* (GTP-C) is partly reused. Figure 4.1 illustrates a GTP' service model.

In this model, the GTP' protocol is built on top of UDP/IP (Figure 4.1(1)). Above the GTP' protocol, a *charging agent* (or CDR sender; Figure 4.1(2)) is implemented in the GSN, and a *charging server* (Figure 4.1(3)) is implemented in the CG. This GTP' service model follows the *primitive flow model* [Lin01], where a GSN communicates

Charging for Mobile All-IP Telecommunications Yi-Bing Lin and Sok-Ian Sou

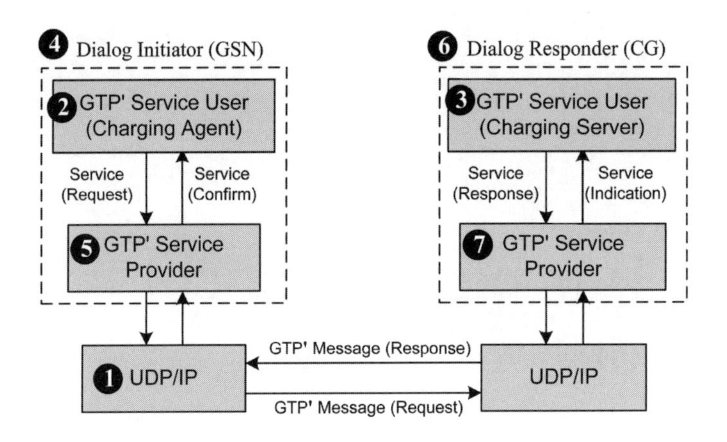

Figure 4.1 The GTP' service model

with a CG through a dialog by invoking the GTP' service primitives. A service primitive can be one of four types: Request (REQ), Indication (IND), Response (RSP), and Confirm (CNF). A service primitive is initiated by a GTP' service user of the dialog initiator. In Figure 4.1, the dialog initiator (Figure 4.1(4)) is a GSN and the service user is a charging agent. The charging agent issues a service primitive with type REQ. This service request is sent to the GTP' service provider of the GSN (Figure 4.1(5)). The service provider sends the request to the dialog responder (Figure 4.1(6)) by creating a GTP' message. This GTP' message is delivered through lower layer protocols; i.e., UDP/IP. When the GTP' service provider of the CG receives the request, it invokes the same service primitive with type IND to the charging server (GTP' service user). The charging server then performs appropriate operations, and invokes the same service primitive with type RSP. This response primitive is a service acknowledgment sent from the CG to the GSN. After the GTP' service provider of the GSN has received this response, it invokes the same service primitive with type CNF. If a dialog is initiated by the CG, then the roles of the CG and the GSN are switched in Figure 4.1.

Based on the above GTP' service model, this section describes the GTP' formats and types of messages exchanged between the peer GTP' service providers (i.e., Figure 4.1(5) and (7)).

4.1.1 The GTP' Message Formats

As defined in 3GPP [3GP02, 3GP05], the GTP' header may follow the standard 20-octet GTP header format (Figure 4.2(a)) or a simplified six-octet format (Figure 4.2(b)). The six-octet GTP' header is the same as the first six octets of the standard GTP header. Octets 7–20 of the GTP header are used to specify a data session between a GSN and the MS. These octets are not needed in GTP'.

Octets	Bits 8	7	6	5	4	3	2	1
1	Version			P	Spare ' 1 1 1'			0
2	Message Type							
3-4	Length							
5-6	Sequence Number							
7-8	Flow Label							
9	SNDCP N-PDULLC Number							
10	Spare ' 1 1 1 1 1 1 1 1 '							
11	Spare ' 1 1 1 1 1 1 1 1 '							
12	Spare ' 1 1 1 1 1 1 1 1 '							
13-20	TID							

(a) GTP header (Version 0)

Octets	Bits 8	7	6	5	4	3	2	1
1	Version			P	Spare ' 1 1 1 '			1
2	Message Type							
3-4	Length							
5-6	Sequence Number							

(b) Six-octet header

Figure 4.2 GTP' header formats

Figure 4.2 shows the GTP' header formats. In this figure, the first bit of octet 1 is used to indicate the header format. If the value is 1, the six-octet header is used. If the value is 0, the 20-octet standard GTP header is used. Note that better GTP' performance is expected by using the six-octet format, because the unused GTP header fields are eliminated. On the other hand, it is easier to support GTP' in an existing GTP environment if the standard GTP header format is used. In commercial deployment, CG/GSN typically accepts both six-octet and 20-octet GTP' headers, but it is recommended to send six-octet GTP' header packets in order to save the network bandwidth. The six-octet header format is described as follows:

- Bit 1 of octet 1 is set to 1 for the six-octet header.
- Bits 2–4 of octet 1 are reserved (with value 111).
- The *Protocol Type* (PT; bit 5 of octet 1) and the *Version* (bits 6–8 of octet 1) fields are used to specify the protocol being used (GTP or GTP' in R99, R4, R5 and so on). For a GTP' message, PT = 0, and the version number can be 000 for GTP' v0.
- The *Message Type* is specified in octet 2. A GTP' message can be Echo, Version Not Supported, Node Alive, Redirection, or Data Record Transfer. Details of these message types will be elaborated later.

- The *Length* field (octets 3 and 4) indicates the payload length. All information following the GTP' header is considered GTP' payload.
- The *Sequence Number* (octets 5 and 6) is used as a transaction identity for the request and response messages. The Sequence Number in a GTP' response message shall be copied from the corresponding request message. Ranging from 0 to 65535, this number unambiguously defines a GTP' request message sent to the peer GTP' node.

Table 4.1 lists the GTP' *Message Types*. Three GTP message types are reused in GTP', including Echo Request, Echo Response and Version Not Supported. A GTP' message contains a GTP' header and the payload. The payload may contain several *Information Elements* (IEs). The IEs used in the GTP' protocol are listed in Table 4.2. Details of the GTP' messages and GTP' IEs are described in the following subsections.

Table 4.1 GTP' message types

Message type	Description
1	Echo Request
2	Echo Response
3	Version Not Supported
4	Node Alive Request
5	Node Alive Response
6	Redirection Request
7	Redirection Response
240	Data Record Transfer Request
241	Data Record Transfer Response

Table 4.2 GTP' information elements

IE type	Description
1	Cause
14	Recovery
126	Packet Transfer Command
249	Sequence Numbers of Released Packets
250	Sequence Numbers of Cancelled Packets
251	Node Address
252	Data Record Packet
253	Requests Responded
254	Address of Recommended Node
255	Private Extension

4.1.2 Echo Message

The Echo Request (Message Type = 1) and Echo Response (Message Type = 2) message pair is typically used to check if the peer is alive. These path management messages are required if GTP' is supported by UDP. Specifically, the Echo Request is sent from a GSN to check if the peer CG is alive. The CG may optionally send the Echo Request to the GSN. An Echo Request message does not include any message payload. On the other hand, an Echo Response message includes a two-octet payload to carry the *Recovery* IE (IE Type value 14). The *Recovery* IE contains the local restart counter value for the CG. Details of the local restart counter are described below.

In 3GPP TS 29.060 [3GP07], the Echo Request is periodically sent for more than 60 seconds on each connection. Whenever a CG receives the Echo Request, it replies with the Echo Response that contains the value of its local restart counter. This counter is maintained in both the GSN and the CG to indicate the number of restarts performed at the CG. If the restart counter value received by the GSN is larger than the value previously stored, the GSN assumes that the CG has restarted since the last Echo Request/Response message pair exchange. In this case, the GSN may retransmit the earlier unacknowledged packets to the CG rather than wait for expiries of their timers. In some commercial implementations, if the GSN does not receive the response message from the CG in four seconds, the GSN will retransmit the request message. Message retransmission is repeated for three times (besides the first sending). Clearly, the above settings may not be optimal for all operation scenarios. Performance evaluation of GTP' timers and retransmission will be discussed in Appendix A.

4.1.3 Version Not Supported Message and Node Alive Message

Version Not Supported (Message Type = 3) indicates the latest GTP version that can be supported by the GTP entity on the identified UDP/IP address. This message is returned when the recipient cannot recognize the version number in the request message. In this message, no payload is included. The sequence number of the Version Not Supported message is the same as that in the previous GTP' message received with unsupported version field.

Node Alive Request (Message Type = 4) and Response (Message Type = 5) message pair is used to inform that a CG has restarted its service after a service break. The service break may be caused by, e.g., node failure or hardware maintenance.

Node Alive Request is sent from a GSN to a CG when the GSN is activated, or when the CG joins the network configuration. The payload of this message is either a six-octet or 18-octet *Node Address* IE (IE Type value 251) indicating the GSN's IP address (either in the IPv4 or the IPv6 format). After the response message is received from the CG, the GSN will not send any Node Alive Request to that CG unless the GSN is restarted. The response message does not include any payload. The GTP' sequence number is reset after this operation.

The CG may optionally send this request message to the GSN to notify that it is online. Therefore, the GSN can learn early that the CG is available. The CG includes its own IP address in the *Node Address* IE. This message has a similar effect as periodic Echo Request.

4.1.4 The Redirection Message

When a CG is not available for service (due to outage for maintenance or external failure) or the when the CG loses connection to the billing system, the CG sends the Redirection Request (Message Type = 6) message to inform a GSN (that originally connects to this CG) to redirect its CDRs to another CG. This message can also be used to balance the workloads among the CGs. The payload of this message includes *Cause* IE (IE Type value 1). The *Cause* value can be "System failure", "Send buffers becoming full", "Receive buffers becoming full", "Another node is about to go down", "This node is about to go down", etc. An *Address of Recommended Node* IE (IE Type value 254) may be given to indicate an alternative CG to take over incoming GSN traffic. If this IE is included, the GSN will start sending messages to the specified IP address.

In response to Redirection Request, the GSN sends the Redirection Response (Message Type = 7) to the CG. This message includes a two-octet *Cause* IE with the value "Request accepted", "No resources available", "Service not supported", "System failure", "Invalid message format", "Version not supported", and so on. Note that the Redirection Request message is never sent from a GSN to a CG.

4.1.5 Data Record Transfer Message

A Data Record Transfer Request/Response message pair is used for CDR delivery. The maximum number of CDRs that the GGSN aggregates in a Data Record Transfer Request message is 255. In this request message (Message Type = 240), the payload includes the *Packet Transfer Command* IE (IE Type value 126). The request may also contain the *Data Record Packet* IE (IE Type value 252), *Sequence Numbers of Released Packets* IE (IE Type value 249) or *Sequence Numbers of Cancelled Packets* IE (IE Type value 250).

If the *Packet Transfer Command* IE indicates "Send Data Record Packet" (code = 1) or "Send possibly duplicated Data Record Packet" (code = 2), the second IE consists of one or more CDRs encapsulated in the *Data Record Packet* IE. Note that in some commercial products, the GSN is only allowed to send one data record at a time. Figure 4.3 shows the format of *Data Record Packet* IE. In this IE, the *Data Record Format* and the *Data Record Format Version* fields are used to identify the CDR format. The CDRs follow the *Abstract Syntax Notation One* (ASN.1) format or other formats as identified by the *Data Record Format* field. The *Data Record Format Version* field identifies the technical specification release and version number that are used for the CDR encoding.

Octet	Bits 8 7 6 5 4 3 2 1
1	Type=252 (Decimal)
2-3	Length
4	Number of Data Records
5	Data Record Format
6-7	Data Record Format Version
8-9	Length of Data Record 1
10...n	Data Record 1
x...x+1	Length of Data Record N
x+2...y	Data Record N

Figure 4.3 Data record packet information element

If the *Packet Transfer Command* IE indicates "Cancel Data Record Packet" (code = 3) or "Release Data Record Packet" (code = 4), the second IE consists of the *Sequence Numbers of Released Packets* IE or *Sequence Numbers of Cancelled Packets* IE. These IEs contain the sequence numbers for the previous Data Record Transfer messages with *Packet Transfer Command* "Send possibly duplicated data record packet". Note that the CG never sends the Data Record Transfer Request message to a GSN. A GSN always waits for a period before it transfers next charging data to the CG. In some commercial GGSN products, the default period is 105 seconds.

In the Data Record Transfer Response (Message Type = 241) message, the header is followed by the *Cause* IE. This IE is a code that indicates how a CDR is processed in the CG. The *Cause* value can be "Request accepted", "CDR decoding error", "Request not fulfilled", "Request already fulfilled", "Request related to possibly duplicated packet already fulfilled", "Sequence numbers of released/cancelled packets IE incorrect" and so on. If the *Cause* value is not "Request accepted", the GSN logs the reason why the request is rejected, treats it as a transmission failure, and retransmits the request. When the CG receives the retransmitted Data Record Transfer Request message, it should respond to each of these packets, but only record it for billing once.

4.2 Connection Setup Procedure

Before a GSN can send CDRs to a CG, a GTP' connection must be established between the charging agent in the GSN and the charging server in the CG. The GTP' connection setup procedure is described in the following steps (see Figure 4.4):

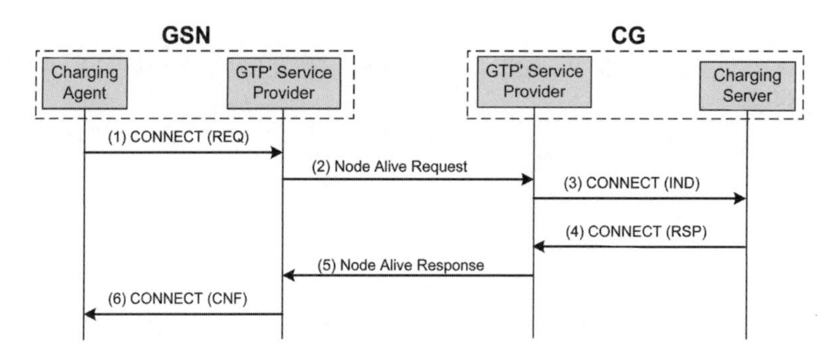

Figure 4.4 Message flow for GTP' connection setup

Step 1. The charging agent instructs the GTP' service provider to set up the connection. This task is initiated by issuing the CONNECT (REQ) primitive with the CG address.

Step 2. The service provider generates a Node Alive Request message and delivers it to the CG through UDP/IP. The UDP source port number is locally allocated at the GSN. On the CG side, the default UDP destination port number is 3386 reserved for GTP' [3GP05]. Alternatively, this destination port number for the CG can be reconfigured.

Step 3. The GTP' service provider of the CG interprets the Node Alive Request message and reports this connection setup event to the charging server via the CONNECT (IND) primitive.

Step 4. The charging server creates and sets a new entry in the GSN list (for this new connection), and responds to the service provider with the CONNECT (RSP) primitive. Either the charging server is ready to receive the CDRs or it is not available for this connection. In the latter case, the charging server may include the address of a recommended CG in the CONNECT (RSP) primitive for further redirection request.

Step 5. Suppose that the CG is available. The GTP' service provider generates a Node Alive Response message, and delivers this message to the GSN.

Step 6. The GTP' service provider of the GSN receives the Node Alive Response message. It interprets the message and reports this acknowledgment event to the charging agent through the CONNECT (CNF) primitive. The charging agent creates and sets the CG entry's status as "active" in the CG list. At this point, the setup procedure is complete.

4.3 CDR Transfer Procedure

The charging agent is responsible for CDR generation in a GSN. The CDRs are encoded using, for example, the ASN.1 format [3GP05]. The charging server is

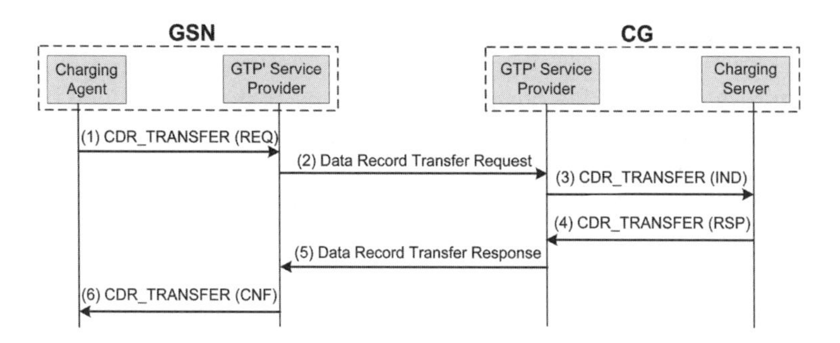

Figure 4.5 GTP' CDR transfer message flow

responsible for decoding the CDRs and returns the processing results to the GSN. The CDR transfer procedure is illustrated in Figure 4.5 and is described in the following steps:

Steps 1. The charging agent encodes the CDR to be transferred. Then it invokes the CDR_TRANSFER (REQ) primitive. This primitive instructs the GTP' service provider to generate a Data Record Transfer Request message.

Steps 2. The service provider includes the CDR in the Data Record Transfer Request message and sends it to the CG.

Steps 3. When the service provider of the CG receives the GTP' message, it issues the CDR_TRANSFER (IND) primitive to inform the charging server that a CDR is received. The charging server decodes the CDR and stores it in the CDR database. This CDR may be consolidated with other CDRs, and is later sent to the billing system.

Steps 4 and 5. The charging server invokes the CDR_TRANSFER (RSP) primitive that requests the GTP' service provider to generate a Data Record Transfer Response message. The *Cause* IE value of the message is "Request accepted". The service provider sends this GTP' message to the GSN.

Steps 6. The GTP' service provider of the GSN receives the Data Record Transfer Response message and reports this acknowledgment event to the charging agent via the CDR_TRANSFER (CNF) primitive. The charging agent deletes the delivered CDR from its unacknowledged buffer.

The primitive flow model described in this section is also used to implement the SS7 protocol. The reader is referred to Chapter 10 in [Lin01] for the details.

4.4 Prepaid Quota Management

GTP' is typically designed for postpaid charging. By extending the GTP' protocol, we show how to design new types of messages for quota management to provide the prepaid service. In the prepaid service, a GSN interacts with a *Prepaid Quota Management* (PQM) server. We introduce several PQM messages based on the GTP' protocol. These new messages can be implemented by using the reserved GTP' message types, or encapsulated in the *Private Extension* IE in GTP' Data Record Transfer Request/Response messages. In this section, we use a *Private Extension* IE to encapsulate a PQM message. As illustrated in Figure 4.6, each *Private Extension* IE contains an Extension Identifier, and an Extension Value. The PQM message type can be carried by the Extension Identifier, and the PQM message attributes can be carried by the Extension Value. We first describe the PQM message types. Then based on these message types, we illustrate the prepaid quota management procedure.

Octet	Bits 8 7 6 5 4 3 2 1
1	Type=252 (Decimal)
2-3	Length
4-5	Extension Identifier
6-m	Extension Value

Figure 4.6 Private extension information element

4.4.1 PQM Message Types

We describe three PQM message types: Authorization Request/Response, Reservation Request/Response, and Quota Reclaim Request/Response. The values for the PQM message types are listed in Table 4.3.

Table 4.3 PQM message types

PQM type	Description
1	Authorization Request
2	Authorization Response
3	Reservation Request
4	Reservation Response
5	Quota Reclaim Request
6	Quota Reclaim Response

The Authorization Request (PQM Message Type = 1) is used for prepaid user authorization. As illustrated in Table 4.4(a), this message contains the *Subscriber*

Identity attribute to identify the user and the *Service Type* attribute to specify the type of the prepaid service. The identity of the subscriber may be the *International Mobile Subscriber Identity* (IMSI) or the *Mobile Station ISDN Number* (MSISDN) of the mobile user. The *Service Type* may be "GPRS session", "SMS" or "MMS". Before the service is authorized, the GSN buffers the user packets and will not forward them to the destination.

The PQM server replies with the Authorization Response (PQM Message Type = 2) when the authorization is finished. As shown in Table 4.4(b), the response contains the *Result* attribute and the *Prepaid Session ID* attribute. The *Result* value may be "Accept" or "Reject". The *Prepaid Session ID* is used to uniquely identify the prepaid session. A *Cause* attribute is included when the *Result* value is "Reject". This *Cause* value may be "Unauthorized user", "Unauthorized service", and so on.

Table 4.4 Attributes contained in the authorization operation

(a) Authorization Request		(b) Authorization Response	
Attribute	Presence requirement	Attribute	Presence requirement
Subscriber Identity	Mandatory	Result	Mandatory
Service Type	Mandatory	Prepaid Session ID	Conditional
		Cause	Conditional

The Reservation Request (PQM Message Type = 3) is sent from the GSN to the PQM server for prepaid credit reservation. It contains several attributes as shown in Table 4.5(a). The *Prepaid Session ID* attribute uniquely identifies a prepaid session which is authorized by the PQM server through the Authorization Request/Response message exchange. The *Request Type* attribute identifies the type of the reservation. For event-based charging, the value is "Event". For session-based charging, the value can be "Initial", "Update" or "Terminate". The *Reporting Reason* attribute specifies the reason for usage. The value can be "Requesting quota", "Quota reclaim requested" or "Validity timer expiry". The reason "Quota reclaim requested", for example, can be used to reclaim credit units from the currently served prepaid sessions. More details on prepaid credit reclaim will be discussed in Appendix E. The *Unit* attribute specifies the non-monetary service unit measured for a particular quota type, e.g., "Second", "KiloByte (KB)", "Byte", "Event", "Packet" and so on. The *Used Quota* attribute contains the amount of used quota. The *Unused Quota* attribute contains the amount of unused quota.

The Reservation Response (PQM Message Type = 4) is sent from the PQM server to the GSN. As illustrated in Table 4.5(b), this message contains the *Granted Quota* attribute and may include the *Validity Timer* attribute to identify the expiry time

Table 4.5 Attributes contained in the reservation operation

(a) Reservation Request		(b) Reservation Response	
Attribute	Presence requirement	Attribute	Presence requirement
Prepaid Session ID	Mandatory	Prepaid Session ID	Mandatory
Request Type	Mandatory	Granted Quota	Mandatory
Reporting Reason	Mandatory	Validity Timer	Optional
Unit	Conditional		
Used Quota	Conditional		
Unused Quota	Conditional		

for the quota. If the *Granted Quota* attribute is 0, it means that the user has insufficient balance or no credit left in the prepaid account.

The Quota Reclaim Request (PQM Message Type = 5) is sent from the PQM server to redistribute quota among various prepaid service sessions of a user when the remaining balance is too low. This message forces the in-progress prepaid sessions to return their unused quota to the PQM server. As shown in Table 4.6, the Quota Reclaim Request contains the *Prepaid Session ID* attribute and the *Cause* attribute. The *Cause* value may be "Not enough balance left in the account" or "Service purged by the system".

Table 4.6 Attributes contained in the quota reclaim request

Attribute	Presence requirement
Prepaid Session ID	Mandatory
Cause	Mandatory

When the GSN receives the Quota Reclaim Request from the PQM server, it acknowledges the PQM server with the Quota Reclaim Response (PQM Message Type = 6). In this response message, no attribute is included. The GSN should immediately report the unused quota by sending the Reservation Request message with *Request Type* "Update" and *Reporting Reason* "Quota reclaim requested". Alternatively, the GSN can reply with the Reservation Request message as a quota reclaim acknowledgment without sending the Quota Reclaim Response (to be elaborated in Steps 3 (a)–(c) of Figure 4.7).

4.5 Prepaid Quota Management Procedure

Consider the scenario where a prepaid user starts a prepaid GPRS session. The quota management procedure is described in the following steps (see Figure 4.7):

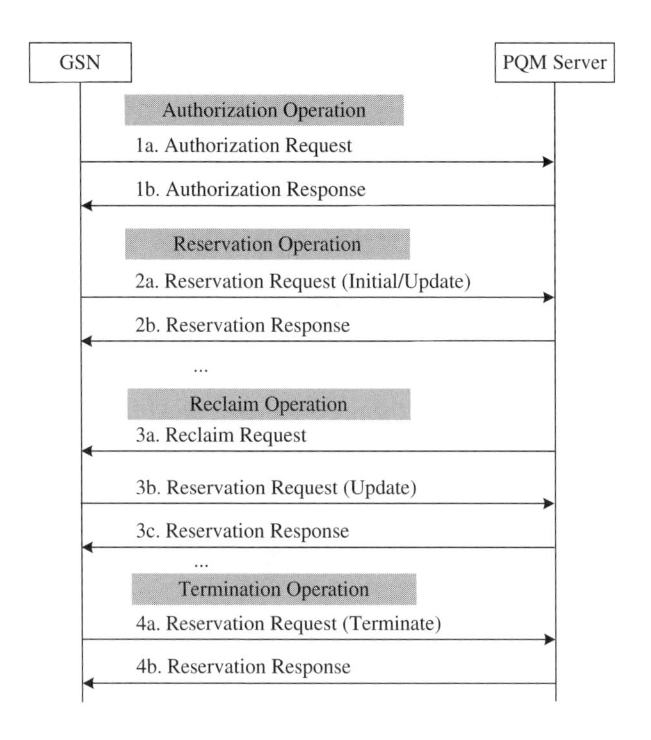

Figure 4.7 Prepaid quota management message flow

Step 1a. To authorize the prepaid session, the GSN sends the Authorization Request message (including the *Subscriber Identity* with the mobile user's IMSI and the *Service Type* with value "GPRS session") to the PQM server.

Step 1b. When the authorization is finished, the PQM server replies the Authorization Response message (which contains the *Result* "Accept" and a unique *Prepaid Session ID*) to the GSN.

Step 2a. The GSN sends the Reservation Request message to the PQM server to reserve the prepaid quota for the service. This message contains the *Request Type* and the *Prepaid Session ID*. The request message is with *Request Type* "Initial" for the first reservation.

Step 2b. Upon receipt of the Reservation Request message, the PQM server determines the tariff of the requested service and then reserves an amount of credit (say, 100 KB). After the reservation is performed, the PQM server acknowledges the GSN with the Reservation Response message including credit reservation information (e.g., *Granted Quota* "100" and *Unit* "KB").

Steps 2a and 2b (with *Request Type* "Update") repeat several times before this service session is complete. If the remaining balance at the PQM server is not large enough to support a new session request (or a served prepaid session), Step 3 may be executed to redistribute quota already granted to the in-progress prepaid sessions.

Step 3a. When the remaining balance at the PQM server is too low, the Quota Reclaim Request message with *Cause* "Not enough balance left in the account" is sent from the PQM server to the GSN. This message forces the in-progress prepaid session to return the unused quota.

Step 3b. When the Quota Reclaim Request message is received, the GSN immediately reports the unused quota by sending the Reservation Request message with *Request Type* "Update" and *Reporting Reason* "Quota reclaim requested". Then the PQM server may redistribute and reserve the prepaid quota among various prepaid sessions. Several policies used for quota redistribution are discussed in Appendix E [Lin06].

Step 3c. After the quota redistribution and the reservation are performed, the PQM server acknowledges the GSN with the Reservation Response message including new credit reservation information.

When a prepaid service session is complete, the following steps are executed:

Step 4a. The GSN sends the Reservation Request message with *Request Type* "Terminate" to the PQM server. This action terminates the session and reports the amount of the consumed credit.

Step 4b. The PQM server releases the unused reserved credit, and acknowledges the GSN with the Reservation Response message.

4.6 Concluding Remarks

In UMTS, the GTP' protocol is used to deliver the CDRs from GSNs to CGs. This chapter presented the message types used in the GTP' protocol. We showed how these messages are used in the connection setup and the CDR transfer procedures. Based on GTP', we also showed how to design new message types for prepaid quota management. In Appendix A, we will show how to support reliability and availability for GTP' connection.

Review Questions

1. GTP' can be implemented by using UDP or TCP. Describe the advantages and disadvantages of these approaches.
2. Describe two kinds of GTP' protocol message headers. What are the tradeoffs between these two formats?

3. Describe the nine messages types used in the GTP' protocol. Which message types are not required if GTP' is implemented by using TCP?
4. With the Redirection Request message, the CGs can be switched. Design a load-balancing algorithm that balances the workload of the CGs using GTP'. What are the criteria for switching the CGs? How does the switching cost affect the switching decision in your algorithm?
5. What message types are required to implement prepaid service? Show how to reclaim the prepaid credit of a user through these messages.
6. Multiple CDRs can be aggregated and delivered in a Data Record Transfer message. However, many commercial implementations only allow one CDR per message. Why?
7. Describe the Prepaid Quota Management (PQM) procedure.
8. Design an authorization algorithm that can be used by the PQM server to authorize the GSN.
9. Describe two different message flows for PQM server to reclaim the quotas from the GSNs.

References

[3GP02] 3GPP, 3rd Generation Partnership Project; Technical Specification Group Core Network; General Packet Radio Service (GPRS); GPRS Tunneling Protocol (GTP) across the Gn and Gp Interface (Release 1998), 3G TS 09.60 version 7.10.0 (2002-12), 2002.

[3GP05] 3GPP, 3rd Generation Partnership Project; Technical Specification Group Services and Systems Aspects; Telecommunication management; Charging management; Charging data description for the Packet Switched (PS) domain (Release 5), 3G TS 32.215 version 5.9.0 (2005-06), 2005.

[3GP07] 3GPP, 3rd Generation Partnership Project; Technical Specification Group Core Network; General Packet Radio Service (GPRS); GPRS Tunneling Protocol (GTP) across the Gn and Gp Interface (Release 6), 3G TS 29.060 version 6.18.0 (2007-09), 2007.

[Lin01] Lin, Y.-B. and Chlamtac, I., *Wireless and Mobile Network Architectures*. John Wiley & Sons Ltd., Chichester, UK, 2001.

[Lin06] Lin, P., Lin, Y.-B., Yen, C.-S., and Jeng, J.-Y., Credit allocation for UMTS prepaid service, *IEEE Transactions on Vehicular Technology*, **55**(1): 306–317, 2006.

5

Mobile Charging Protocols

In the previous chapter, we introduced the GTP' charging protocol for GPRS offline charging. Besides GTP', protocols such as *Customized Applications for Mobile network Enhanced Logic* (CAMEL), *Remote Authentication Dial In User Service* (RADIUS) and Diameter are widely deployed for charging in telecommunications networks. This chapter describes how these protocols are used in mobile networks. We first introduce CAMEL that was primarily designed for traditional circuit-switched services. Then we introduce RADIUS and Diameter charging protocols that are utilized in next generation all-IP networks. Finally, we describe the charging related fields that are included in the *Session Initiation Protocol* (SIP) for the IMS applications. With these fields, the SIP interacts with the Diameter protocol to accommodate IMS charging.

5.1 Customized Application for the Mobile Network Enhanced Logic (CAMEL)

Intelligent Network (IN) allows fixed or mobile operators to offer enhanced telephony services such as number translation and prepaid call. As an IN standard, CAMEL provides control intelligence across a mobile communications network. CAMEL has been evolved in four phases. CAMEL phases 1 and 2 were developed by the *European Telecommunications Standards Institute* (ETSI) [3GP00]. Phase 1 was tailored for the GSM-based core networks. Phase 2 extended phase 1 with a greater range of options such as prepaid call charging. In this phase, a pre-recorded announcement can be played to alert the prepaid user of a prepaid call when the user's credit is depleted. CAMEL phases 3 and 4 are developed by 3GPP [3GP06d]. Phase 3 supports control capabilities for mobile services including SMS and GPRS. CAMEL phase 4

Charging for Mobile All-IP Telecommunications Yi-Bing Lin and Sok-Ian Sou

is extensible for any enhancements. In particular, it provides control capabilities for the IMS services.

Figure 5.1 shows the CAMEL functional architecture [3GP00, 3GP05a]. The *GSM Service Switching Function* (gsmSSF; Figure 5.1(a)) is implemented in a switching node such as *Mobile Switching Center* (MSC; Figure 5.1(g)). The *GSM Service Control Function* (gsmSCF; Figure 5.1(b)) implements operator-specific services through the CAMEL service logic. The *GSM Specialized Resource Function* (gsmSRF; Figure 5.1(c)) provides various resources (such as voice prompt, announcements and DTMF digit collection) that can be allocated to support interaction between users and the gsmSRF. A standalone gsmSRF is implemented in an IN node called the intelligent peripheral (Figure 5.1(d)). The intelligent peripheral communicates with the MSC through the *ISDN User Part* (ISUP) protocol. The *CAMEL Application Part* (CAP) is a signaling protocol used to support the information flows between CAMEL functional entities including gsmSSF, gsmSCF and gsmSRF.

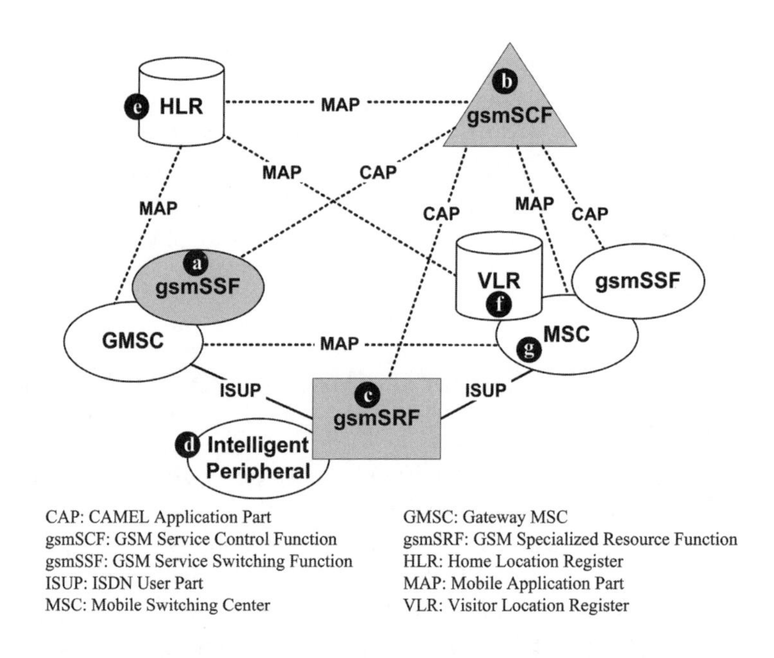

Figure 5.1 CAMEL functional architecture

In this architecture, the *Home Location Register* (HLR; Figure 5.1(e)) stores the *CAMEL Subscription Information* (CSI) for subscribers requiring CAMEL support. Several CSIs implemented in CAMEL are listed in Table 5.1. Every CSI type corresponds to a CAMEL service. For example, an *Originating CSI* (O-CSI) indicates that a user has subscribed to an originating CAMEL service. The O-CSI includes the following attributes:

- The *gsmSCF address* attribute specifies the E.164 address of the gsmSCF where the CAMEL service logic resides.
- The *service key* attribute identifies the CAMEL service in the gsmSCF. Examples are prepaid service and 0800 toll-free service.
- A *trigger Detection Point* (DP) attribute indicates a specific event in the gsmSSF. This event initiates a dialogue to the gsmSCF. For example, DP Collected_Info informs the gsmSCF that the dialed digits have been collected; DP Answer indicates that the call is answered; DP Disconnect indicates that the call is terminated. When a DP is encountered, the call processing is suspended and the gsmSSF notifies the gsmSCF to handle the call control.
- The *default call handling* attribute describes which action the gsmSSF should perform in case of signaling failure between the gsmSSF and the gsmSCF. For example, when the gsmSSF fails to communicate with the gsmSCF, the gsmSSF should terminate the call processing if the default call handling is "release".

Table 5.1 Types of CAMEL subscription information

CSI	Description
O-CSI	Information for invoking CAMEL service logic in a mobile originating call
T-CSI	Information for invoking CAMEL service logic in a mobile terminating call
OSMS-CSI	Information for invoking CAMEL service logic in a mobile originating SMS submission
TSMS-CSI	Information for invoking CAMEL service logic in a mobile terminating SMS delivery
GPRS-CSI	Information for invoking CAMEL service logic in a GPRS session

For an MS with originating CAMEL service, the O-CSI is downloaded from the HLR to the *Visitor Location Register* (VLR; Figure 5.1(f)) at location update. GSM location update is achieved through *Mobile Application Part* (MAP) message exchanges. The VLR stores the O-CSI as a part of the subscriber data for a mobile user roaming in this VLR area. When processing an originating call for the MS, the MSC will receive an O-CSI from the VLR. During call processing, the MSC is requested to monitor the call states and to inform the gsmSSF that some actions must be taken based on the trigger DPs (e.g., DP Answer, DP Disconnect and so on). When the event is detected, the gsmSSF suspends call processing and sends a request to the gsmSCF. The gsmSCF will reply with an instruction such as "continue", "release" or "connect". For example, the gsmSSF continues the prepaid call establishment if the subscriber's balance stored in the gsmSCF is sufficient (and it therefore receives the "continue" instruction from the gsmSCF); the gsmSSF may release the prepaid call if the subscriber's balance is depleted; the gsmSSF may connect to a specified called party (e.g., the telephone number for a local hospital or a local police station) based on the called number and the location. Details of prepaid call service implemented by CAMEL will be discussed in Section 6.1 and Appendix E. Details of the MAP and the ISUP protocols can be found in [Lin01].

5.2 Remote Authentication Dial In User Service (RADIUS)

The *Remote Authentication Dial In User Service* (RADIUS) protocol was originally defined by the *Internet Engineering Task Force* (IETF) to provide a centralized *Authentication, Authorization and Accounting* (AAA) framework for network access [IET00a, IET00b]. RADIUS is developed based on the client–server architecture, and is commonly used in *Network Access Servers* (NASs) such as wireless access points and VoIP gateways. An example of VoIP billing solution using the RADIUS protocol will be described in Chapter 10. In this protocol, the RADIUS client (an NAS; see Figure 5.2(a)) is responsible for passing user information to the RADIUS server (Figure 5.2(b)), and then takes some actions based on the returned response. The RADIUS server receives user connection requests, authenticates the user, and then returns all configuration information necessary to the client for service delivery.

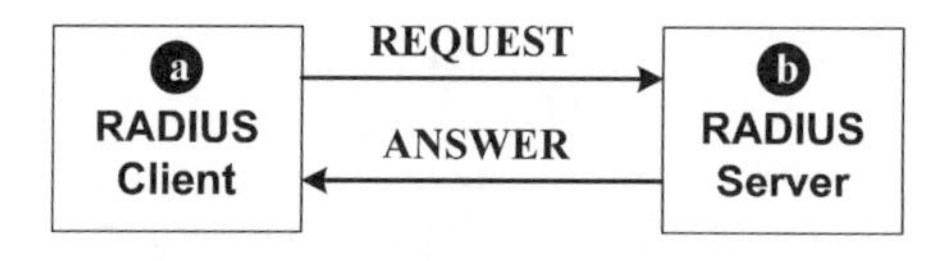

Figure 5.2 The RADIUS architecture

A RADIUS message consists of a 20-octet header (see Table 5.2) followed by several attributes. The RADIUS attributes convey information between RADIUS clients and RADIUS servers. Details of the header format are described as follows:

Table 5.2 The RADIUS header format

Octets	
1	Code
2	Identifier
3-4	Length
5-20	Authenticator

- The *Code* field identifies the message type. Examples of RADIUS codes are listed in Table 5.3.
- The *Identifier* field is used to match a request/response message pair.
- The *Length* field indicates the length of the message including the header and the attribute fields. The minimum length is 20 octets and the maximum length is 4096 octets.
- The *Authenticator* field is used to authenticate the messages between the RADIUS client and the RADIUS server.

Table 5.3 The RADIUS codes

Code	Message name	Description
1	Access-Request	To determine whether a user is allowed to access a specific service
2	Access-Accept	To provide specific configuration information for service delivery
3	Access-Reject	To reject a service request when the Access-Request message format is wrong, the password is incorrect, and so on
4	Accounting-Request	To provide accounting information for a service delivery
5	Accounting-Response	To acknowledge an Accounting-Request
255	Reserved	—

RADIUS attributes carry specific AAA information described as follows:

- The *User-Name* attribute specifies the name of the user.
- The *User-Password* attribute indicates the password of the user.
- The *NAS-IP-Address* attribute contains the IP address of the network access server (i.e., the RADIUS client).
- The *Service-Type* attribute specifies the type of requested service.
- The *Acct-Status-Type* attribute shows the status of an Accounting-Request message. The attribute value can be Start (value 1), Interim-Update (value 3) or Stop (value 2).
- The *Acct-Input-Octets* attribute indicates how many octets have been received by the network access server.
- The *Acct-Output-Octets* attribute indicates how many octets have been sent out from the network access server.
- The *Acct-Session-Id* attribute records a unique accounting identity that matches the start and the stop accounting records.
- The *Acct-Session-Time* attribute indicates the elapsed time of the service session.
- The *Acct-Terminate-Cause* attribute indicates why the session is terminated, which can only be presented in the Accounting-Request message where the

Acct-Status-Type is set to "Stop". This attribute can be User Request (value 1), Session Timeout (value 5), or Admin Reboot (value 7).
- The *Called-Station-Id* attribute specifies the called phone number.
- The *Calling-Station-Id* attribute specifies the calling phone number.

5.3 Diameter

The Diameter protocol was derived from RADIUS to offer more flexibility, and is the next generation AAA protocol [IET03a]. Diameter is an extensible messaging protocol enabling AAA within and across IP multimedia networks. The Diameter protocol has fail-over capabilities, runs over, for example, secure TCP/SCTP transport. Its modular architecture offers a flexible base protocol which allows application-specific extensions. Diameter has proven successful in overcoming the limitations of RADIUS. Therefore, rapid growth in the usage of Diameter-based charging can be expected. The 3GPP has chosen the Diameter protocol to enable AAA capabilities for the IMS network [RAD07, 3GP06b].

Like RADIUS, Diameter follows the client–server architecture where a client (Figure 5.3(1)) and a server (Figure 5.3(2)) interact through the Diameter request and answer message exchange. A Diameter relay agent (Figure 5.3(3)) may be used to forward a Diameter message to the appropriate destination.

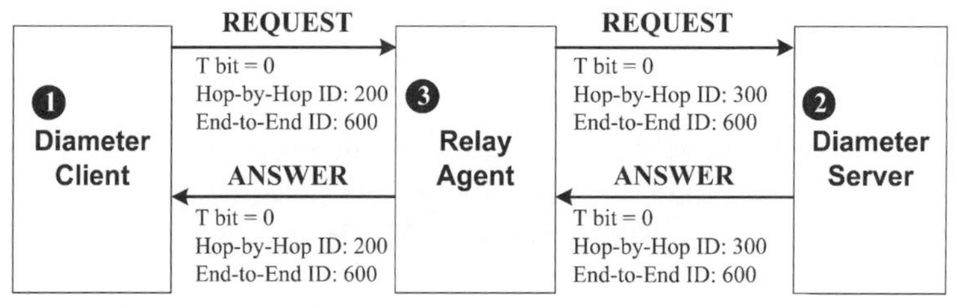

(a) Diameter message transmission

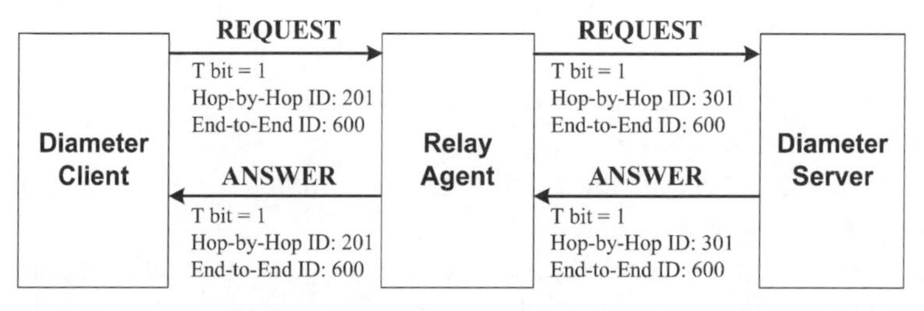

(b) Diameter message retransmission

Figure 5.3 Diameter client and server

Several Diameter applications defined by the IETF are utilized in IMS, including the Diameter Credit Control Application for IP-based online charging control [IET05, 3GP07a]. Depending on the release of 3GPP, the Diameter accounting servers are implemented in different network nodes. In 3GPP R5, the offline charging information is collected by the *Charging Collection Function* (CCF) to be described in Chapter 7. The CCF transfers charging information from the IMS nodes to the network operator's billing system without going through a *Charging Gateway Function* (CGF). In 3GPP R6, the offline charging information is collected by the *Charging Data Function* (CDF), and the online charging information is collected by CGF implemented in the *Online Charging System* (OCS). The details will be described in Chapter 8. The CDF receives charging information from the IMS nodes or the GPRS support nodes, and uses the charging information to construct well-defined CDRs. These CDRs are transferred to the CGF and then to the billing system.

Table 5.4 The Diameter header format

Octets	0	1	2	3	4	5	6	7
1	Version							
2–4	Message Length							
5	R	P	E	T	Reserved			
6–8	Command Code							
9–12	Application-ID							
13–16	Hop-by-Hop Identifier							
17–20	End-to-End Identifier							

A Diameter message consists of a 20-octet header (see Table 5.4) followed by several *Attribute-Value Pairs* (AVPs). The Diameter header fields are described as follows:

- The *Version* field is set to "1".
- The *Message Length* field indicates the length of the Diameter message.
- The *Command Flags* field contains eight bits. Four bits (Bits 1–4) are reserved. The "R" bit (Bit 0) indicates that the message is a request (with value 1) or an answer (with value 0). The "P" bit (Bit 1) indicates that the message is proxied (with value 1) or locally processed (with value 0). The "E" bit (Bit 2) is always set to value 0 in a request message. In an answer message, this bit indicates that there is an error in the request message (with value 1) or no error (with value 0). The "T" bit (Bit 3) indicates that the message is the original message (with value 0; see Figure 5.3(a)) or a retransmitted message (with value 1; see Figure 5.3(b)).
- The *Command Code* field identifies the message type. Each request/answer message pair is assigned a unique command code [3GP07a, 3GP07d, 3GP07e]. Some examples of the command codes for Diameter charging messages are listed in Table 5.5.

- The *Application-ID* field identifies the target application of the message. Examples of Diameter charging application IDs are listed in Table 5.6.
- The *Hop-by-Hop Identifier* is used to match a Diameter answer message with the corresponding request message. In a request message, the Hop-by-Hop Identifier value is replaced at each hop as the Diameter request message is relayed to its final destination. In Figure 5.3(a), when the Diameter request message is forwarded from the relay agent to the Diameter server, the Hop-by-Hop Identifier value 200 is replaced by 300 in the relay agent.
- The *End-to-End Identifier* is a sequence number used to detect a duplicate Diameter message. The End-to-End Identifier is not modified when a Diameter message is forwarded to its final destination. In Figure 5.3, the End-to-End Identifiers are filled with the value 600 in both the original transmission and the retransmission.

Table 5.5 Diameter command codes

Command code	Message name	Abbreviation
258	Re-Auth Request / Answer	RAR/RAA
271	Accounting Request / Answer	ACR/ACA
272	Credit Control Request / Answer	CCR/CCA
274	Abort Session Request / Answer	ASR/ASA
275	Session Termination Request / Answer	STR/STA

Table 5.6 Diameter application IDs

Application ID	Application Name	Comment
0	Common Messages	—
3	Accounting	This ID is also specified in the *Acct-Application-Id* AVP of the ACR/ACA messages described in Section 5.4.1
4	Credit Control	This ID is also specified in the *Auth-Application-Id* AVP of the CCR/CCA messages described in Section 5.5.1

5.4 Diameter-based Offline Charging

Offline charging for both events and sessions are performed between an IMS node and the CDF through the Rf reference point (see Figure 5.4) which is designed for non-real-time operations. The CDF functionalities will be elaborated in Chapter 8. In this section, we describe the message formats and the procedures for IMS offline charging.

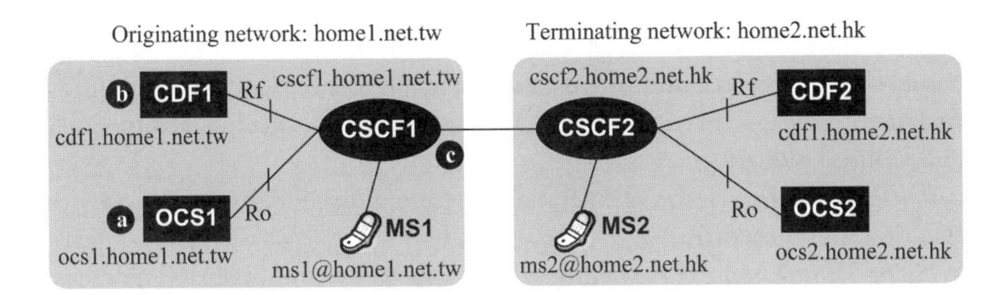

Figure 5.4 An example of IMS charging configuration (where the calling party is MS1 and the called party is MS2)

5.4.1 Offline Charging Message Formats

IMS offline charging is achieved by exchanging the Diameter Accounting Request (ACR) and Accounting Answer (ACA) messages. The ACR message contains the following AVPs [IET03a, 3GP07a]:

- The *Session-Id* AVP identifies the offline charging session.
- The *Origin-Host* and the *Origin-Realm* AVPs contain the address and the realm for the Diameter client. As shown in Figure 5.4, if an IMS session is established from a user MS1 to another user MS2, the Origin-Host and Origin-Realm will be cscf1.home1.net.tw and home1.net.tw, respectively.
- The *Destination-Realm* AVP contains the realm of the Diameter server. The Destination-Realm format is the same as that for the Origin-Realm. In Figure 5.4, the Destination-Realm is home2.net.hk.
- The *Acct-Application-Id* AVP contains the value "3" as defined in RFC 3588 [IET03a]. This value indicates the Diameter accounting application (see Table 5.6).
- The *Accounting-Record-Type* AVP indicates the *transfer type* of the offline accounting information. There are four transfer types: EVENT_RECORD (value 1) indicates that an event-based service is delivered. An event-based service creates exactly one EVENT_RECORD record (to be elaborated in Figure 5.5). START_RECORD (value 2) initiates a charging session (typically when the IMS node receives a SIP 200 OK acknowledging a SIP INVITE message). This record contains charging information relevant to the initiation of the session. INTERIM_RECORD (value 3) contains up-to-date charging information for an existing charging session (e.g., when a specified interim interval has elapsed for the SIP session). STOP_RECORD (value 4) terminates a charging session (when the IMS node receives a SIP BYE message). This record contains final accounting information of the session. A session-based service consists of one START_RECORD, one STOP_RECORD and zero or more INTERIM_RECORD records (to be elaborated in Figure 5.6).

- The *Accounting-Record-Number* AVP identifies the record within a session.
- The *User-Name* AVP contains the user name in the *Network Access Identifier* (NAI) format according to RFC 3588 [IET03a]. In Figure 5.4, the User-Name of MS1 is ms1@home1.net.tw.
- The *Acct-Interim-Interval* AVP specifies the time interval between the previous and the next charging records.
- The *Event-Timestamp* AVP specifies the time when the accounting message is created.
- The *Service-Information* AVP contains the service-specific parameters. For GPRS service, this AVP includes the GPRS charging identifier, the PDP type/address, the SGSN and the GGSN addresses, and so on. For the IMS service, this AVP includes the calling and the called party addresses, the application server information, the IMS charging identifier, and the IMS media information. For the WLAN service (see Figure 2.12), this AVP includes the WLAN session identifier, the *Packet Data Gateway* (PDG) charging identifier, the PDG and the *Wireless Access Gateway* (WAG) addresses. For the MMS service, this AVP includes the originator and the recipient addresses, the submission time and the message ID.

The ACA message contains the following AVPs:

- The *Result-Code* AVP contains the result of the accounting request. In Table 5.7 at Section 5.5.1, the result code DIAMETER_SUCCESS (value 2001) indicates that the service request is successfully completed. The result code DIAMETER_TOO_BUSY (value 3004) indicates that the Diameter server is busy.
- The *Session-Id*, the *Origin-Host*, the *Origin-Realm*, the *Acct-Application-Id*, the *Accounting-Record-Type*, the *Accounting-Record-Number*, the *User-Name*, the *Acct-Interim-Interval*, and the *Event-Timestamp* AVPs are the same as those in the ACR message.

5.4.2 Offline Charging Procedures

In order to exchange charging information, the ACR message is sent from a Diameter client (e.g., CSCF1 in Figure 5.4) to a Diameter server (e.g., CDF1 in Figure 5.4) through the Rf interface. The Diameter server then sends the ACA message to acknowledge an ACR message. In event-based charging, the network node reports the service usage through the ACR message with *Accounting-Record-Type* "EVENT_RECORD". Figure 5.5 shows the Diameter offline message flow for the event-based service between a network node (e.g., an IMS application server) and the CDF, which is described in the following two steps:

Step 1. The network node sends the ACR message to the CDF. This message reports the charging information such as the event timestamp, the user name, the network node address, and the service information. The *Accounting-Record-Type* field in the ACR message is set to "EVENT_RECORD".

Step 2. Upon receipt of the ACR message, the CDF processes the offline charging information and produces a CDR. Then through the ACA message, the CDF acknowledges the network node that the charging information has been received.

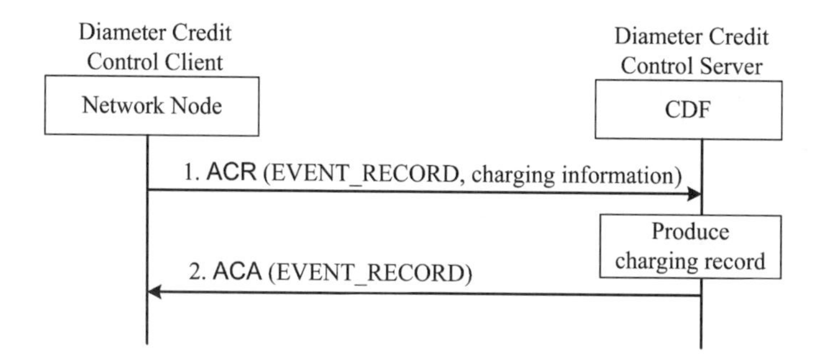

Figure 5.5 Message flow for event-based offline charging

In the session-based offline charging, the charging information is reported through the procedure described below (see Figure 5.6).

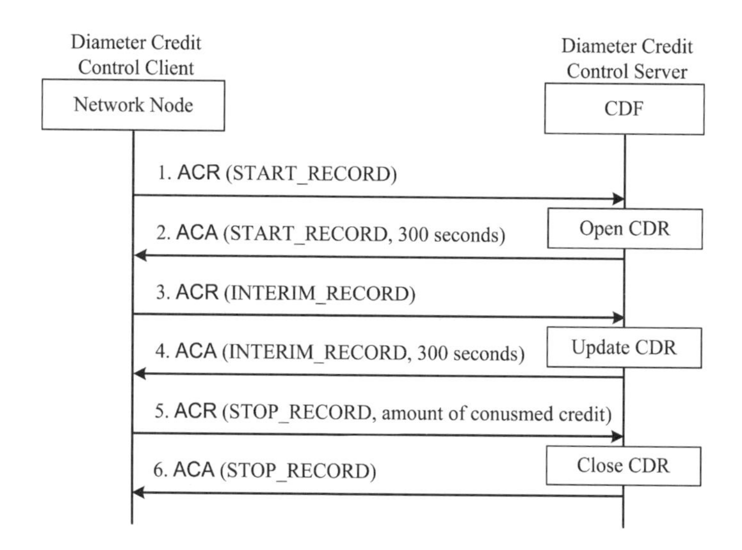

Figure 5.6 Message flow for session-based offline charging

Step 1. The network node (e.g., GGSN or S-CSCF) sends the ACR message with *Accounting-Record-Type* "START_RECORD" to the CDF.

Step 2. After the CDF has opened a CDR for the offline session-based service, it sends the ACA message to the network node to indicate that the charging information is received. In the ACA message, an *Acct-Interim-Interval* value (e.g., 300 seconds) can be included to instruct when the next ACR message should be sent. If the *Acct-Interim-Interval* value is not included, Steps 3 and 4 are skipped.

Step 3. Assume that the *Acct-Interim-Interval* value is contained in the ACA message. When the interim interval has elapsed, the network node sends the up-to-date charging information to the CDF by using the ACR message with *Accounting-Record-Type* "INTERIM_RECORD".

Step 4. The CDF updates the CDR and acknowledges the network node with the ACA message. Steps 3 and 4 may repeat many times before the service session terminates.

Step 5. When the service session is complete, the network node sends the ACR message with *Accounting-Record-Type* "STOP_RECORD" to the CDF. This action terminates the session and reports the final charging information (e.g. the total credit units consumed by the requested service).

Step 6. The CDF updates and closes the CDR. Then it sends the ACA message to the network node to indicate that the session is terminated.

5.5 Diameter-based Online Charging

The all-IP mobile network utilizes the *Diameter Credit Control* (DCC) application to communicate with the OCS through the Ro interface (see Figure 5.4) [3GP06a, 3GP07a]. In the packet-switched service domain [3GP07b], online charging for GPRS session involves the GGSN (i.e., the DDC client) and the OCS (i.e., the DCC server). The GGSN uses the Credit Control Request (CCR) message to transfer the GPRS charging information to the OCS. The OCS uses the Credit Control Answer (CCA) message to reserve credit units (or quotas) for the GPRS session. The charging information collected by the GGSN includes the duration of the served GPRS session and the transferred data volume. Details of charging for a GPRS session will be described in Section 6.2. In IMS online charging, the IMS network node (e.g., CSCF) acts as a DCC client and the OCS acts as a DCC server. This section describes the message formats and the procedures for IMS online charging.

5.5.1 Online Charging Message Formats

The CCR message is sent from the DCC client to the DCC server to request an amount of credit for an online service. The CCR message contains the following AVPs [IET05, 3GP07a]:

- The *Session-Id* AVP identifies the credit control session.
- The *Origin-Host* and the *Origin-Realm* AVPs contain the address and the realm of the DCC client. In Figure 5.4, the *Origin-Host* is cscf1.home1.net.tw. The *Origin-Realm* is home1.net.tw.
- The *Destination-Host* and the *Destination-Realm* AVPs contain the address and the realm of the DCC server. In Figure 5.4, the *Destination-Host* is ocs1.home1.net.tw. The *Destination-Realm* is home1.net.tw.
- The *Auth-Application-Id* AVP contains value "4" as defined in RFC 4006 [IET05]. This value indicates the Diameter Credit Control Application (see Table 5.6).
- The *Service-Context-Id* AVP contains the identifier allocated by the service provider or by the standardization body. The format of the *Service-Context-Id* is "extensions.MNC.MCC.Release.service-context@domain". In 3GPP R6, the identifier of the PS charging is "MNC.MCC.6.32251@3gpp.org". The identifier of the WLAN charging is "MNC.MCC.6.32252@3gpp.org", and the identifier of the IMS charging is "MNC.MCC.6.32260@3gpp.org".
- The *CC-Request-Type* AVP indicates the credit control request type for the online service. A request message for credit control can be one of the following types: INITIAL_REQUEST (value 1) initiates a credit control session. UPDATE_REQUEST (value 2) contains update credit control information for an in-progress session. This request is sent when the credit units currently allocated for the session are completely consumed. TERMINATION_REQUEST (value 3) terminates an in-progress credit control session. EVENT_REQUEST (value 4) is used in one-time credit control for event-based service.
- The *CC-Request-Number* AVP contains the sequence number of the message.
- The *Subscription-Id* AVP contains the identification of the user, which can be a SIP *Uniform Resource Identifier* (URI), an IMSI or an MSISDN.
- The *Termination-Cause* AVP is presented if the *CC-Request-Type* is set to TERMINATION_REQUEST. This AVP provides the reason why the credit control session is terminated. For example, DIAMETER_LOGOUT (value 1) indicates that the DCC client terminates the credit control session.
- The *Requested-Action* AVP is only presented in the EVENT_REQUEST message. This AVP specifies the action for the event, such as DIRECT_DEBITING, REFUND_ACCOUNT, CHECK_BALANCE or PRICE_ENQUIRY.
- The *Multiple-Services-Credit-Control* AVP contains the parameters for quota management, including the amount of request credit, the amount of used credit, the reporting reason, the identity of the used service, and the identifier of the rating group.
- The *User-Name*, the *Event-Timestamp*, the *Service-Information* AVPs are similar to those in the ACR message.

The CCA message is sent from the DCC server to the DCC client to grant an amount of credit for the online service. The CCA contains the following AVPs:

- The *Result-Code* AVP contains the result of the credit control request. Examples of the result codes are listed in Table 5.7 [IET05].
- The *CC-Session-Failover* AVP contains an indication to inform the DCC client whether or not a fail-over handling is used.
- The *Credit-Control-Failure-Handling* AVP determines the action in the Diameter client when the Diameter server is temporarily not available, e.g., because of network failure. The action may be TERMINATE (with value 0), CONTINUE (with value 1) or RETRY_AND_TERMINATE (with value 2).
- The *Multiple-Services-Credit-Control* AVP contains the parameters for the quota management, including the amount of granted credit, the identifier of the requested service, the identifier for the rating group, the validity time for the usage of granted credit units (i.e., the time when the DCC client should perform another credit control request), and the events (such as QoS changes or SGSN changes) that will trigger the credit reauthorization procedure.
- The *Cost-Information* AVP contains the cost information of the requested service.
- The *Session-Id*, the *Origin-Host*, the *Origin-Realm*, the *Auth-Application-Id*, the *CC-Request-Type*, the *CC-Request-Number*, the *Service-Information* AVPs are similar to those described in the CCR message.

Table 5.7 Examples of result codes in the Diameter answer message

Result Code	Value	Description
DIAMETER_SUCCESS	2001	The request is successfully completed
DIAMETER_TOO_BUSY	3004	The Diameter server is busy
DIAMETER_END_USER_SERVICE_DENIED	4010	The service request is denied
DIAMETER_CREDIT_LIMIT_REACHED	4012	The service request is denied because the remaining credit units are not sufficient for the requested service
DIAMETER_USER_UNKNOWN	5030	The end user is not found in the DCC server
DIAMETER_RATING_FAILED	5031	The DCC server fails to rate the service request

5.5.2 Online Charging Procedures

There are three kinds of online charging: immediate event charging, event charging with unit reservation, and session charging with unit reservation. This subsection describes the message flows of these procedures in detail.

In the credit control process for immediate event charging, the unit credits are immediately deducted from the user account in one single transaction. The message flow is shown in Figure 5.7 with the following steps:

Step 1. [Debit Units (request)] When the network node (e.g., IMS application server) receives an event-based online service request, the Debit Units operation is performed prior to service delivery. The network node sends the CCR message with *CC-Request-Type* "EVENT_REQUEST" to the OCS. The *Requested-Action* field is set to "DIRECT_DEBITING".

Step 2. [Debit Units (response)] Upon receipt of the CCR message, the OCS determines the price of the requested service and then debits an equivalent amount of credit. The OCS acknowledges the network node with the CCA message to authorize the service delivery. This CCA message includes the granted credit units and the cost information of the requested service. Then the network node starts to deliver the event-based service.

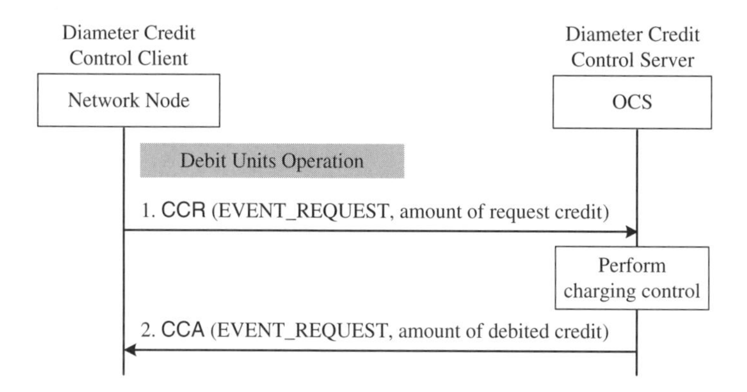

Figure 5.7 Message flow for immediate event charging

There are two basic operations in the aforementioned charging control:

- *Unit determination* refers to the calculation of the number of non-monetary units (service units, data volume, time and events) that shall be assigned prior to starting service delivery.
- *Rating* refers to the calculation of a price out of the non-monetary units provided by the unit determination function.

Both unit determination and rating are performed at the OCS in Step 2 of Figure 5.7. Alternatively, unit determination can be conducted at the network node, and the result is included in the CCR message in Step 1 of Figure 5.7. Based on the amount of units indicated in the CCR message, the OCS performs the rating function to compute the price for the service at Step 2. As a third alternative, both unit determination and rating can be performed at the network node. The derived monetary units are then included in the CCR message at Step 1. At Step 2, the OCS exercises the accounting control and authorizes the requested monetary units.

Immediate event charging is adopted when the price for an event-based service is clearly defined. Consider a scenario where a user is charged with $5 for viewing a movie clip. In this example, the OCS directly deducts $5 from the user's account and authorizes the IMS network nodes (such as Media Resource Function Controller) to provide the movie clip to the user. In Section 8.2.1, we will give a more detailed example of immediate event charging for messaging service. In this kind of charging, the network node must ensure that the requested service delivery is successful. If the service delivery fails, the network node should send another CCR message with *CC-Request-Type* "EVENT_REQUEST" and *Requested-Action* "REFUND_ACCOUNT" to the OCS. The OCS then refunds the credit units that are previously debited.

To avoid failure handling for event-based service, it is more appropriate to use event charging with unit reservation as shown in Figure 5.8, which is described in the following steps:

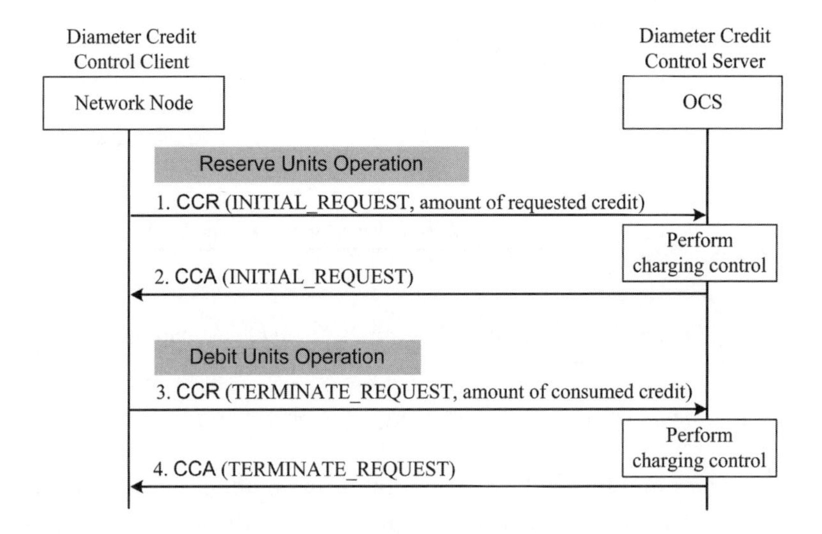

Figure 5.8 Message flow for event charging with unit reservation

Step 1. [Reserve Units (request)] The network node (e.g., IMS application server) sends the CCR message with *CC-Request-Type* "INITIAL_REQUEST" to the OCS. This message indicates the amount of requested credit.

Step 2. [Reserve Units (response)] Upon receipt of the CCR message, the OCS determines the price of the requested service and then reserves an equivalent amount of credit. Once the reservation has been made, the OCS replies with the CCA message to authorize the service delivery to the network node. The network node starts to deliver the service.

Step 3. [Debit Units (request)] When the service delivery is complete, the network node sends the CCR message with *CC-Request-Type* "TERMINATION_REQUEST" to the OCS. This message terminates the credit control session and reports the amount of the consumed credit.

Step 4. [Debit Units (response)] The OCS debits the consumed credit units from the subscriber's account and releases the unused reserved credit units. The OCS acknowledges the network node with the CCA message. This message contains the cost information of the requested service.

In Section 8.2.2, we will give a more detailed example for IMS event charging with unit reservation.

In the session charging with unit reservation, the unit credits are reserved by the OCS from the user's account mainly because, in this situation, the OCS does not know how many units are needed to provide the service. Upon the service termination, the amount of used credit is deducted from the user's account, and eventually any units reserved and not used are released and added back to the user's account. The credit reservation procedure for session-based online charging includes three types of credit control operations: Reserve Units operation (Steps 1 and 2 in Figure 5.9), Reserve Units and Debit Units operation (Steps 3 and 4 in Figure 5.9) and Debit Units operation (Steps 5 and 6 in Figure 5.9). The procedure is described as follows:

Step 1. [Reserve Units (request)] To start the service with credit reservation, the network node (e.g., GGSN or S-CSCF) sends the CCR message with *CC-Request-Type* "INITIAL_REQUEST" to the OCS. This message indicates the amount of requested credit.

Step 2. [Reserve Units (response)] Upon receipt of the CCR message, the OCS determines the tariff of the requested service and then reserves an equivalent amount of credit. After the reservation is performed, the OCS acknowledges the network node with the CCA message including credit reservation information.

Step 3. [Reserve Units and Debit Units (request)] During the service session, the granted credit units may be depleted. If so, the network node sends a CCR message with *CC-Request-Type* "UPDATE_REQUEST" to the OCS. Through this message, the network node reports the amount of used credit, and requests for additional credit units.

Step 4. [Reserve Units and Debit Units (response)] When the OCS receives the CCR message, it debits the amount of consumed credit and reserves extra credit for the service session. The OCS acknowledges the network node by sending the CCA message with the *Result-Code* "DIAMETER_SUCCESS" and the

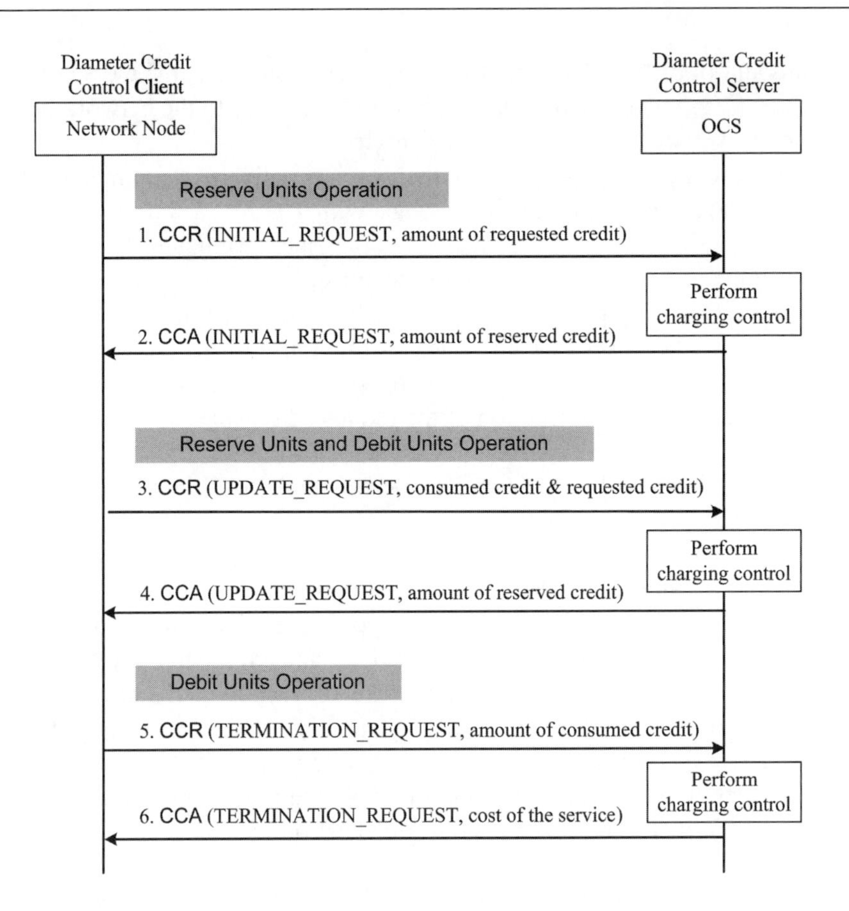

Figure 5.9 Message flow for session charging with unit reservation

amount of the reserved credit. Note that the Reserve Units and Debit Units operation (i.e., Steps 3 and 4) may repeat several times before the service session is complete.

Step 5. [Debit Units (request)] When the service session is complete, the network node sends the CCR message with *CC-Request-Type* "TERMINATION_REQUEST" to the OCS. This action terminates the session and reports the amount of the consumed credit.

Step 6. [Debit Units (response)] The OCS debits the consumed credit units and releases the unused reserved credit units. The OCS acknowledges the network node with the CCA message. This message may contain the total cost of the service.

In Section 8.2.3, we will illustrate a detailed IMS call setup example through session charging with unit reservation.

5.6 Session Initiation Protocol: IMS Charging Headers

The IMS architecture described in Section 2.4 provides real-time multimedia service by utilizing the *Session Initiation Protocol* (SIP) [IET02a, 3GP06b]. In order to correlate the CDRs generated by different IMS nodes and the GGSN for an IMS call session (to be elaborated in Section 7.2), some charging information headers must be included in a SIP message. With these headers, each involved IMS node can obtain the correlated charging data and then include them in the CDRs. In this section, we introduce two SIP headers used for IMS charging.

5.6.1 P-Charging-Function-Addresses

The P-Charging-Function-Address header indicates the address of the charging function which is responsible for collecting the CDRs generated by the IMS nodes. In online charging, the P-Charging-Function-Address header contains the OCS address (e.g., ocs1.home1.net.tw in Figure 5.4(a)). In offline charging, the charging function address is a CCF address in 3GPP R5 or a CDF address in 3GPP R6 (e.g., cdf1.home1.net.tw in Figure 5.4(b)). The offline and online charging function addresses transferred in SIP signaling are encoded in the P-Charging-Function-Address field as defined in 3GPP TS 24.299 [3GP07c] and RFC 3455 [IET03b].

A CSCF (either P-CSCF or S-CSCF; see Figure 5.4(c)) includes the SIP header P-Charging-Function-Addresses in either the initial request or the response for a call session. One or more charging function addresses are passed from the CSCF to the involved IMS nodes in the SIP signaling path. If multiple charging addresses are included in the P-Charging-Function-Addresses header of a SIP message, the first address is the primary address and the secondary address is for redundancy. In Figure 5.4, the P-Charging-Function-Addresses included in the SIP request issued from MS1 is:

```
P-Charging-Function-Addresses:   cdf=cdf1.home1.net.tw;
                                 ocs=ocs1.home1.net.tw;
```

5.6.2 P-Charging-Vector

The IMS charging correlation information is encoded in the P-Charging-Vector header of a SIP message. This header is defined in RFC 3455 [IET03b]. Extensions to this header and its usage within the IMS network are described in 3GPP TS 24.299 [3GP07c]. The P-Charging-Vector header includes the following identifiers.

- *IMS charging identifier* (ICID) identifies an IMS call session (or event service) in the charging/billing system. The charging data generated in different IMS nodes

are correlated with the ICID shared among IMS network nodes involved in the IMS session. This identity is globally unique across all 3GPP IMS networks for a time period of at least one month. Expiration of an ICID can be detected by using IMS node-specific information, e.g., high-granularity time information and location information. At each SIP session establishment, a new ICID is generated at the first IMS node (e.g., P-CSCF) that processes the SIP INVITE message. This ICID is then used in all subsequent SIP messages for that session (e.g., SIP 200 OK and BYE messages) until the session is terminated. The ICID is generated by an IMS node in the UTF8String format (e.g., 1234bc9876e) [3GP06c].
- *Access network charging identifier* specifies the corresponding IP-based bearer that transports the IMS media flows on the originating or the terminating side. This information is used to correlate the access network charging data with the IMS charging data. Examples for access network charging identifiers are *GPRS Charging Identifier* (GCID) for GPRS and PDG charging identifier for WLAN services [3GP05b, 3GP06e]. At PDP context activation, the GCID is generated by the GGSN in the UTF8String format [3GP06c].
- *Inter-operator identifier* (IOI) identifies both originating and terminating networks involved in a SIP session. This identifier is inserted by the S-CSCF and will be removed by the P-CSCF when the SIP request goes outside the IMS network.

In a SIP call establishment, the P-CSCF is typically the first IMS node that processes the SIP request. The P-CSCF includes the P-Charging-Vector header in the SIP request. In Figure 5.4, the P-Charging-Vector for a SIP request issued from MS1 is

```
P-Charging-Vector:    icid-value = 1234bc9876e;
                      gcid = 3069882;
                      orig-ioi = home1.net.tw;
                      term-ioi = home2.net.hk
```

5.7 Concluding Remarks

This chapter described how CAMEL, RADIUS and Diameter are used to implement charging functionalities in mobile telecommunications networks. We introduced the CAMEL functional architecture and the CAMEL subscription information. We briefly described the RADIUS protocol, and then elaborated on how the Diameter charging protocol is utilized in next generation all-IP networks. Specifically, we described the Diameter message formats and the message flows for offline and online charging. At the end of this chapter, we also described the SIP headers that are used for IMS charging.

Several commercial Diameter software tools are available for developing IMS charging. An example is the RADVISION DIAMETER Toolkit for the development of

IMS Diameter-compliant network elements [RAD07]. Designed to ensure maximum flexibility, this toolkit is equipped with API hooks that allow the application layer to control call flows and objects inside the toolkit for enhanced application capabilities. Implemented in the C language, this toolkit also allows multiple applications to run over it transparently.

Another example is MARBEN Diameter suitable for use in all IMS equipment needing Diameter [MAR07]. In this tool, access to the Rf and Ro interfaces as well as the Diameter Base Protocol is offered through a set of dedicated high-level object-oriented APIs in languages such as C++ or Java. This tool supports extensibility for creation of new AVPs, security by processing of Diameter security AVP, and provision for supporting IPsec and *Transport Layer Security* (TLS) secure transport layers. It runs on top of any TCP or *Stream Control Transmission Protocol* (SCTP) layer [IET02b] offering a socket transport interface with very little work. Similarly, the operation and management interface has been kept simple while providing the functionalities expected for telecom software that runs under strict operational constraints.

An example of a Diameter implementation for Rf/Ro interfaces in SIP application server will be given in Appendix F.

Review Questions

1. Describe gsmSSF, gsmSCF and gsmSRF. How do they interact with each other?
2. Give two examples of mobile services that can be implemented by CAMEL.
3. Describe the CSI types and their usage.
4. Describe the RADIUS header format.
5. Describe the Acct-Status-Type and Acct-Terminate-Cause attributes used in the RADIUS request message.
6. What is the major difference between CCF and CDF?
7. Describe the Diameter header format.
8. Describe the accounting record types for the Diameter accounting messages.
9. Describe the Acct-Interim-Interval AVP used in offline charging. Is there any similar AVP used in online charging? What is the advantage and disadvantage of using this kind of AVP?
10. Describe the message flow for the session-based offline charging.
11. Describe the credit control request types for the Diameter credit control messages.
12. Describe the detailed steps of immediate event charging where the unit determination is performed at the network node.

13. What is the major difference between the immediate event charging and event charging with unit reservation?
14. Describe the detailed steps of event charging with unit reservation where both unit determination and rating are performed at the network node.
15. Describe the message flow for the session-based online charging.
16. What is P-Charging-Function-Address? What is P-Charging-Vector?
17. What are ICID and IOI?
18. What is an access network charging identifier? Give two examples.

References

[3GP00] Digital cellular telecommunications system (Phase 2+); Customised Applications for Mobile network Enhanced Logic (CAMEL); CAMEL Application Part (CAP) specification (Release 1997), GSM 09.78 version 6.5.0, (2000-07), 2000.

[3GP05a] 3GPP, 3rd Generation Partnership Project; Technical Specification Group Services and System Aspects; Customized Applications for Mobile network Enhanced Logic (CAMEL); Service description; Stage 1 (Release 5), 3G TS 22.078 version 5.15.0 (2005-03), 2005.

[3GP05b] 3GPP, 3rd Generation Partnership Project; Technical Specification Group Services and Systems Aspects; General Packet Radio Service (GPRS); Service description; Stage 2 (Release 5), 3G TS 23.060 version 5.13.0 (2006-12), 2006.

[3GP06a] 3GPP, 3rd Generation Partnership Project; Technical Specification Group Service and System Aspects; Telecommunication management; Charging management; Online Charging System (OCS): Applications and interfaces (Release 6), 3G TS 32.296 version 6.3.0 (2006-09), 2006.

[3GP06b] 3GPP, 3rd Generation Partnership Project; Technical Specification Group Core Network; IP Multimedia Subsystem (IMS); Stage 2 (Release 5), 3G TS 23.228 version 5.15.0 (2006-06), 2006.

[3GP06c] 3GPP, 3GPP. 3rd Generation Partnership Project; Technical Specification Group Service and System Aspects; Telecommunication management; Charging management; Charging data description for the IP Multimedia Subsystem (IMS) (Release 5), 3G TS 32.225 version 5.11.0 (2006-03), 2006.

[3GP06d] 3GPP, 3rd Generation Partnership Project; Technical Specification Group Core Network; Customised Applications for Mobile network Enhanced Logic (CAMEL) Phase 4; CAMEL Application Part (CAP) specification (Release 6), 3G TS 29.078 version 6.5.0 (2006-06), 2006.

[3GP06e] 3GPP, 3rd Generation Partnership Project; Technical Specification Group Service and System Aspects; Telecommunication management; Charging management; Wireless Local Area Network (WLAN) charging (Release 6), 3G TS 32.252 version 6.2.0 (2006-09), 2006.

[3GP07a] 3GPP, 3rd Generation Partnership Project; Technical Specification Group Service and System Aspects; Telecommunication management; Charging management;

Diameter charging applications (Release 6), 3G TS 32.299 version 6.12.0 (2007-09), 2007.

[3GP07b] 3GPP, 3rd Generation Partnership Project; Technical Specification Group Services and System Aspects; Telecommunication management; Charging management; Packet Switched (PS) domain charging (Release 6), 3GPP TS 32.251 version 6.10.0 (2007-06), 2007.

[3GP07c] 3GPP, 3rd Generation Partnership Project; Technical Specification Group Core Network and Terminals; IP Multimedia Call Control Protocol based on Session Initiation Protocol (SIP) and Session Description Protocol (SDP); Stage 3 (Release 6), 3GPP TS 24.229 version 6.17.0 (2007-12), 2007.

[3GP07d] 3GPP, 3rd Generation Partnership Project; Technical Specification Group Core Network and Terminals; Rx Interface and Rx/Gx signalling flows (Release 6), 3G TS 29.211 version 6.3.0 (2005-12), 2005.

[3GP07e] 3GPP, 3rd Generation Partnership Project; Technical Specification Group Core Network and Terminals; Policy control over Gq interface (Release 6), 3G TS 29.209 version 6.6.0 (2006-09), 2006.

[IET00a] IETF, Remote Authentication Dial In User Service (RADIUS). IETF RFC 2865, 2000.

[IET00b] IETF, RADIUS Accounting. IETF RFC 2866, 2000.

[IET02a] IETF, SIP: Session Initiation Protocol. IETF RFC 3261, 2002.

[IET02b] IETF, An Introduction to the Stream Control Transmission Protocol (SCTP). IETF RFC 3286, 2002.

[IET03a] IETF, Diameter Base Protocol. IETF RFC 3588, 2003.

[IET03b] IETF, Private Header (P-Header) Extensions to the Session Initiation Protocol (SIP) for the 3rd Generation Partnership Project (3GPP). IETF RFC 3455, 2003.

[IET05] IETF, Diameter Credit-Control Application. IETF RFC 4006, 2005.

[Lin01] Lin, Y.-B. and Chlamtac, I., *Wireless and Mobile Network Architectures*. John Wiley & Sons Ltd., Chichester, UK, 2001.

[MAR07] MARBEN Diameter Product, MARBEAN, 2007.

[RAD07] IMS DIAMETER Toolkit. RADVISION, 2007.

6

UMTS CS/PS Charging Management

A traditional mobile telecommunications system provides offline charging where the *Charging Data Records* (CDRs) are collected and then sent to the billing system after a service is delivered. That is, no charging information is sent to the billing system during the service session. A user typically receives a monthly bill that shows the chargeable items during a specific period. Advanced mobile telecom introduces data applications with real-time control and management, which requires a convergent and flexible online charging system. To address this issue, UMTS Release 5 (R5) defines the charging functionalities that allow the network nodes to generate the CDRs accurately. In UMTS R5, online charging furnishes real-time charging information in order to dynamically perform credit control for network resource usage.

In the 3GPP *Circuit Switched* (CS) service domain, the *Mobile Switching Center* (MSC; Figure 6.1(a)) server sends the CDRs to the billing system for offline charging. The CS online charging implements two functions through *Customized Applications for Mobile network Enhanced Logic* (CAMEL) [3GP04, 3GP05e, 3GP06]. In the CAMEL model described in Chapter 5, the *GSM Service Control Function* (gsmSCF; Figure 6.1(b)) performs operator-specific services, and then transfers the CDRs directly to the billing system. The *GSM Service Switching Function* (gsmSSF; Figure 6.1(c)) interfaces the MSC with the gsmSCF. The gsmSSF is typically implemented in the MSC.

In the *Packet Switched* (PS) service domain, the CDRs generated by the *Serving GPRS Support Node* (SGSN; Figure 6.1(d)) and the *Gateway GPRS Support Node*

Charging for Mobile All-IP Telecommunications Yi-Bing Lin and Sok-Ian Sou

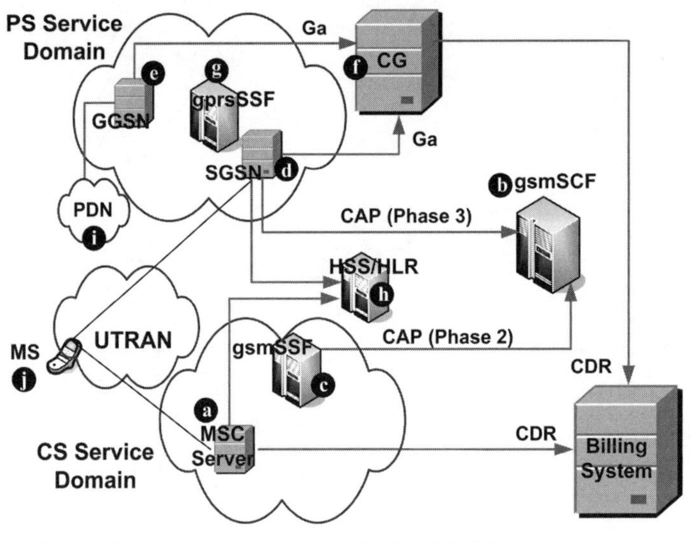

CG: Charging Gateway
GGSN: Gateway GPRS Support Node
HSS: Home Subscriber Server
MS: Mobile Station
PDN: Packet Data Network
SGSN: Serving GPRS Support Node
SSF: Service Switching Function
CS: Circuit Switched
HLR: Home Location Register
MGW: Media Gateway
MSC: Mobile Switching Center
PS: Packet Switched
SCF: Service Control Function
UTRAN: UMTS Terrestrial Radio Access Network

Figure 6.1 Charging for CS and PS service domains

(GGSN; Figure 6.1(e)) are first sent to the *Charging Gateway* (CG; Figure 6.1(f)) for storage and then forwarded to the billing system for offline processing. The CG is responsible for persistent storage of the CDRs and some processing and error checking. The PS online charging is supported by gprsSSF (collocated with SGSN; Figure 6.1(g) and gsmSCF. In this chapter, several charging scenarios are given to explain how the CDRs are generated in the UMTS CS and PS service domains.

6.1 Circuit Switched Service Domain

GSM (2G) and UMTS (3G) support CS based services including voice, CS data and *Short Message Service* (SMS). In the CS service domain, the MSC server and the *Gateway MSC* (GMSC) server are responsible for collecting the charging related information. Through the originating MSC server, all signaling information in a call can be collected even if an inter-MSC handoff occurs (i.e., a call party moves across different MSCs during a call session) [Lin01]. An incoming GMSC is the MSC that can query the *Home Location Register* (HLR; Figure 6.1(h)) to determine the current location of the called *Mobile Station* (MS; Figure 6.1(j)). An outgoing GMSC sets up a call from a GSM network to the *Public Switched Telephone Network* (PSTN) or other networks.

Figures 6.2 and 6.3 illustrate an offline charging example where MS1 in GSM network 1 makes a CS call to MS2 in GSM network 2 [Lin01, 3GP05a, 3GP05b]. It consists of the following steps:

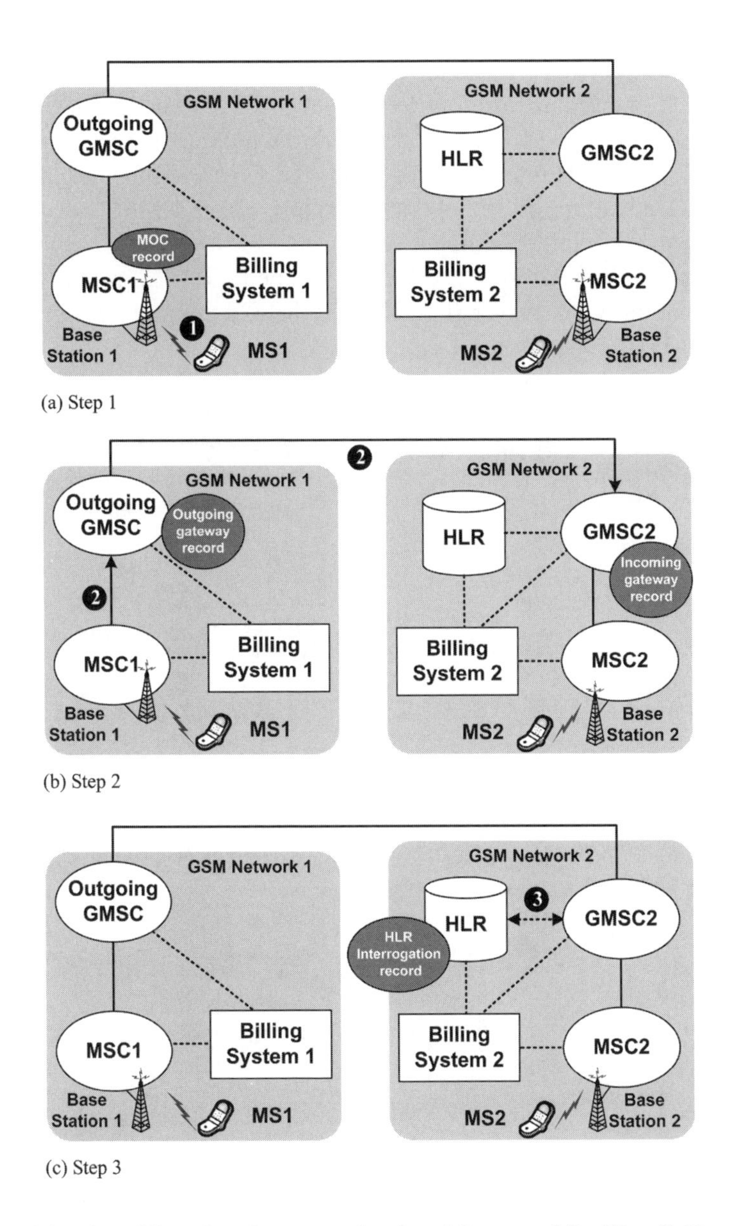

(a) Step 1

(b) Step 2

(c) Step 3

Figure 6.2 An offline charging example of mobile-to-mobile CS call (Steps 1–3)

Step 1. **[Figure 6.2(a)]** Upon receipt of the call setup request from MS1, the originating MSC server (i.e., MSC1) creates a *Mobile Originated Call* (MOC) record for this call.

Step 2. **[Figure 6.2(b)]** The call is routed to the outgoing GMSC of network 1, and then to the GMSC of network 2 (i.e., GMSC2). An outgoing gateway record is created in the outgoing GMSC and an incoming gateway record is created in GMSC2.

Step 3. **[Figure 6.2(c)]** GMSC2 interrogates the HLR to determine the current location (i.e., the *Mobile Station Roaming Number* or MSRN) of MS2. MSRN is a temporary network identity assigned to a mobile subscriber during the call establishment. This number identifies the E.164 number of the terminating MSC. Depending on the operator's setup, either GMSC2 or the HLR may create an HLR interrogation record.

Step 4. **[Figure 6.3(a)]** After the MSRN is obtained, GMSC2 routes the call to the terminating MSC server (i.e., MSC2). MSC2 creates a *Mobile Terminated Call* (MTC) record for this call.

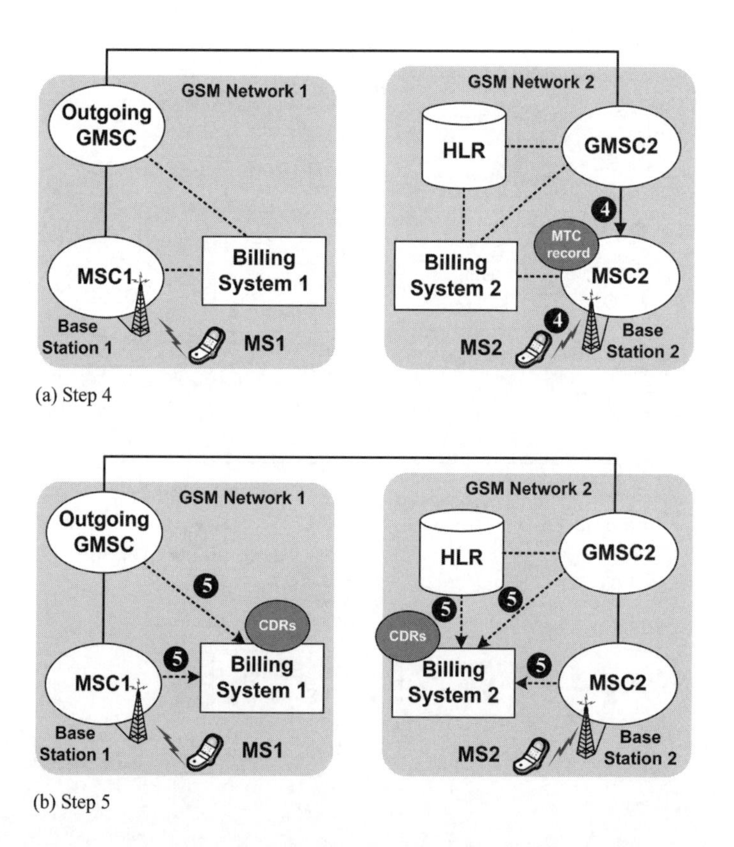

Figure 6.3 An offline charging example of mobile-to-mobile CS Call (Steps 4 and 5)

Step 5. [Figure 6.3(b)] When the call is finished, five CDRs, including the MOC, the MTC, the incoming gateway, the outgoing gateway and the HLR interrogation records, are transferred to the billing systems 1 and 2 via, for example, FTP. The billing systems then process these CDRs for the charging purposes.

In Taiwan, only the originating party will be charged for a mobile phone call (except for international calls), and only the MOC record is processed for billing. In the US and China, both the calling and the called parties are charged. Therefore, both the MOC and the MTC records should be processed for billing. The outgoing/incoming gateway records and the MTC record are used for settlement of accounts with mobile operators involved in roaming traffic. The HLR interrogation record may not be required for billing and settlement, but it may be used for other purposes such as statistical analysis.

An MOC record (generated by the originating MSC in Figure 6.2(a)) contains the following fields [3GP05b]:

- The *Record Type* field indicates "Mobile originated".
- The *Served IMSI* and the *MSISDN* fields specify the *International Mobile Subscriber Identity* (IMSI) and the *Mobile Station ISDN Number* (MSISDN; i.e., the telephone number) of the calling party (i.e., MS1 in Figure 6.2).
- The *Called Number* field specifies the MSISDN of the called party (MS2 in Figure 6.2); i.e., the number dialed by the calling party.
- The *Recording Entity* field identifies the E.164 number of the originating MSC (i.e., MSC1 in Figure 6.2).
- The *Location* field indicates the identity of the cell (i.e., Base Station 1 in Figure 6.2) or *Service Area Code* (SAC) at the time of CDR creation. The SAC is a code of fixed length (two octets) used to identify a service area (for location-based services) within a location area.
- The *Basic Service* field specifies the bearer of the teleservice employed, e.g., telephony (Teleservice No 11), emergency calls (Teleservice No 12), short message mobile terminated (Teleservice No 21), and so on [3GP02].
- The *Call Duration* field indicates the chargeable duration of the connection.
- The *Cause for Termination* field specifies the reason for connection release.
- The *Call Reference* field specifies a local identifier used to distinguish simultaneous transactions for the same MS.

An MTC record (generated by the terminating MSC in Figure 6.3(a)) contains the following fields:

- The *Record Type* field indicates "Mobile terminated".
- The *Served IMSI* and the *MSISDN* fields specify the IMSI and the MSISDN of the called party (MS2).

- The *Calling Number* field specifies the MSISDN of the calling party (MS1). This field is only presented when the calling number is available.
- The *Recording Entity* field identifies the E.164 number of the terminating MSC (i.e., MSC2 in Figure 6.3).
- The *Location* field indicates the identity of the cell (i.e., Base Station 2 in Figure 6.3) or SAC at the time of CDR creation. This field is only presented when the location is available.
- The *Basic Service*, the *Call Duration*, the *Cause for Termination* and the *Call Reference* fields are similar to those in the MOC record.

An outgoing gateway record (generated by the outgoing GMSC in Figure 6.2(b)) contains the following fields:

- The *Record Type* field indicates "Outgoing gateway record".
- The *Called Number* field specifies the MSISDN of the called party (MS2). This number is used by the outgoing GMSC for routing.
- The *Recording Entity* field identifies the E.164 number of the outgoing GMSC.
- The *Outgoing Trunk Group* (TKGP) field indicates the trunk group on which the call leaves the outgoing GMSC.
- The *Event Time Stamp* fields specify the seizure time and the release time of the outgoing trunk. If the call is successfully established, the answer time for the call is also recorded.
- The *Call Duration* field indicates the call holding time of the outgoing trunk.
- The *Cause for Termination* and the *Call Reference* fields are similar to those in the MOC record.

An incoming gateway record (generated by the GMSC2 in Figure 6.2(b)) contains the following fields:

- The *Record Type* field indicates "Incoming gateway record".
- The *Recording Entity* field identifies the E.164 number of the incoming gateway (i.e., GMSC2 in Figure 6.2).
- The *Incoming TKGP* field indicates the trunk group on which the call terminates at GMSC2.
- The *Call Duration* field indicates the call holding time of the incoming trunk.
- The *Called Number*, the *Recording Entity*, the *Event Time Stamp*, the *Cause for Termination*, and the *Call Reference* fields are similar to those in the outgoing gateway record.

An HLR interrogation record (generated by either the HLR or the GMSC in Figure 6.2(c)) contains the following fields:

- The *Record Type* field indicates "HLR interrogation".
- The *Served IMSI* and *MSISDN* fields specify the IMSI and the MSISDN of the party being interrogated (i.e., MS2). The IMSI is only presented if the interrogation is successful.
- The *Recording Entity* field specifies the E.164 number of the HLR/GMSC producing the record.
- The *Interrogation Time Stamp* field specifies the time when the interrogation occurs.

In the CS service domain, online charging (e.g., for prepaid call service) is implemented by using CAMEL. (Appendix E will describe a CAMEL architecture for GPRS online charging.) For a prepaid subscriber, the HLR stores the *Originating CAMEL Subscription Information* (O-CSI), which includes information such as the gsmSCF address, the prepaid service key and a list of the trigger *Detection Points* (DPs) such as DP Collected_Info and DP Disconnect. The HLR passes the subscriber's O-CSI to the *Visitor Location Register* (VLR) on location update. Based on the concept of O-CSI, we show how CAMEL works. The details are explained in the following steps (see Figure 6.4):

Step 1. **[Figure 6.4(a)]** A prepaid call is issued from a mobile user (MS1) to a PSTN user (Phone 2). When the call is set up to the originating MSC (i.e., MSC1), this MSC queries the VLR using MS1's IMSI, and receives an O-CSI from the VLR record.

Step 2. **[Figure 6.4(a)]** Based on the O-CSI information, MSC1/gsmSSF suspends the call and notifies the gsmSCF to handle the call control when DP Collected_Info is encountered. The gsmSSF passes the CAMEL service key (which indicates the prepaid service) to the gsmSCF to request an instruction (e.g., "continue" or "release") for the prepaid call handling. The gsmSCF retrieves the subscriber's profile, applies appropriate tariffs, and converts the account balance into the time units for the call. If the subscriber's balance is depleted,

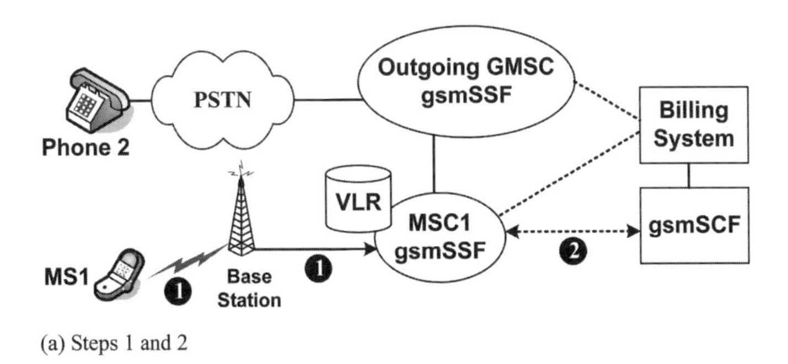

Figure 6.4 Online charging example for prepaid call service

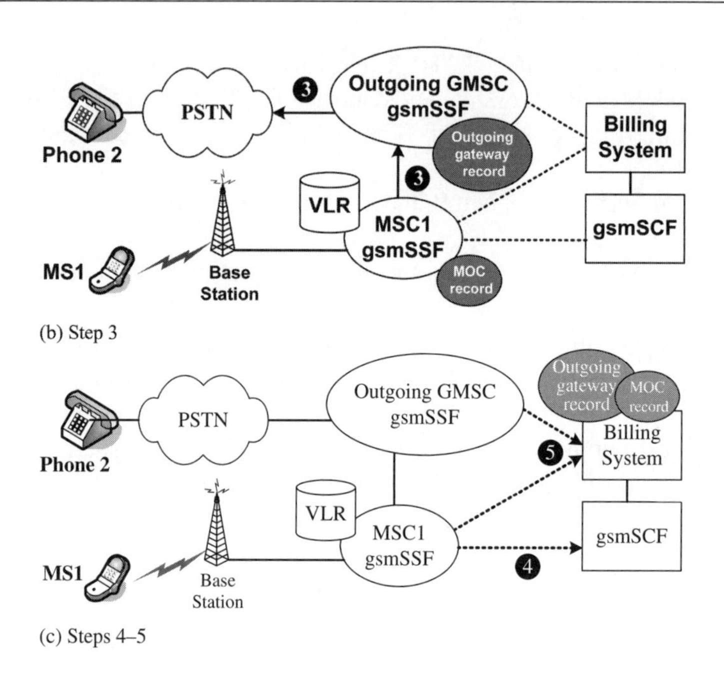

(b) Step 3

(c) Steps 4–5

Figure 6.4 *(continued)*

the gsmSCF returns the instruction "release" to terminate the prepaid call. If the subscriber's balance stored in the gsmSCF is sufficient, the gsmSCF returns the instruction "continue" to authorize the MSC1/gsmSSF to continue the call setup with the calculated call duration time.

Step 3. **[Figure 6.4(b)]** MSC1 then generates an MOC record for MS1, and routes the call to Phone 2 through the outgoing GMSC. The outgoing GMSC server creates an outgoing gateway record for accounting with the PSTN.

Step 4. **[Figure 6.4(c)]** When the prepaid call is terminated, the DP Disconnect is encountered. Then MSC1/gsmSSF notifies the gsmSCF to update the prepaid account balance.

Step 5. **[Figure 6.4(c)]** The generated records, including the MOC and the outgoing gateway records, are subsequently transferred to the billing system when the call is finished.

An MOC record (generated by the originating MSC in Figure 6.4(b)) contains the following fields:

- The *Record Type*, the *Served IMSI*, the *Served MSISDN*, the *Called Number*, the *Recording Entity*, the *Location*, the *Basic Service*, the *Call Duration*, the *Cause for Termination*, and the *Call Reference* fields are similar to those for the MOC record in the example of Figure 6.2.

- The *gsmSCF Address* field identifies the CAMEL server (i.e., the gsmSCF in Figure 6.4) serving the subscriber.
- The *Network Call Reference* field is used for correlation of call records in different network nodes involving in the call session.
- The *MSC Address* field contains the E.164 number assigned to the MSC that generates the network call reference.

An outgoing gateway record (generated by the GMSC in Figure 6.4(b)) is the same as that in the example of Figure 6.2, and the details are omitted.

Figures 6.5 and 6.6 illustrate an example for SMS delivery with the following steps:

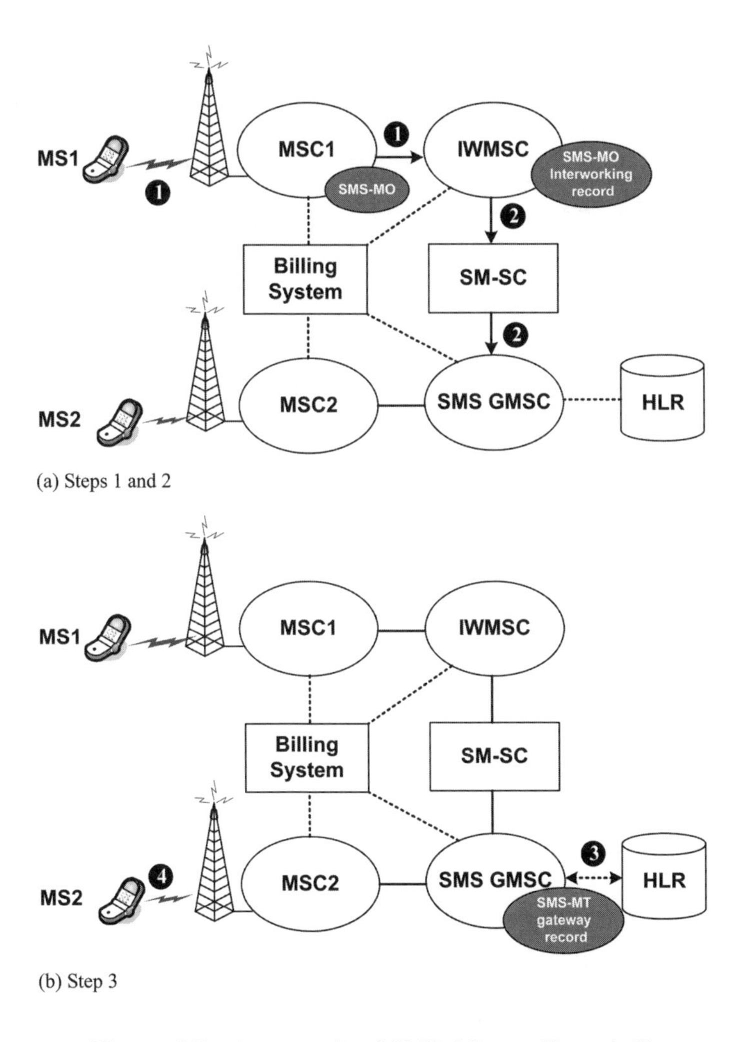

Figure 6.5 An example of SMS delivery (Steps 1–3)

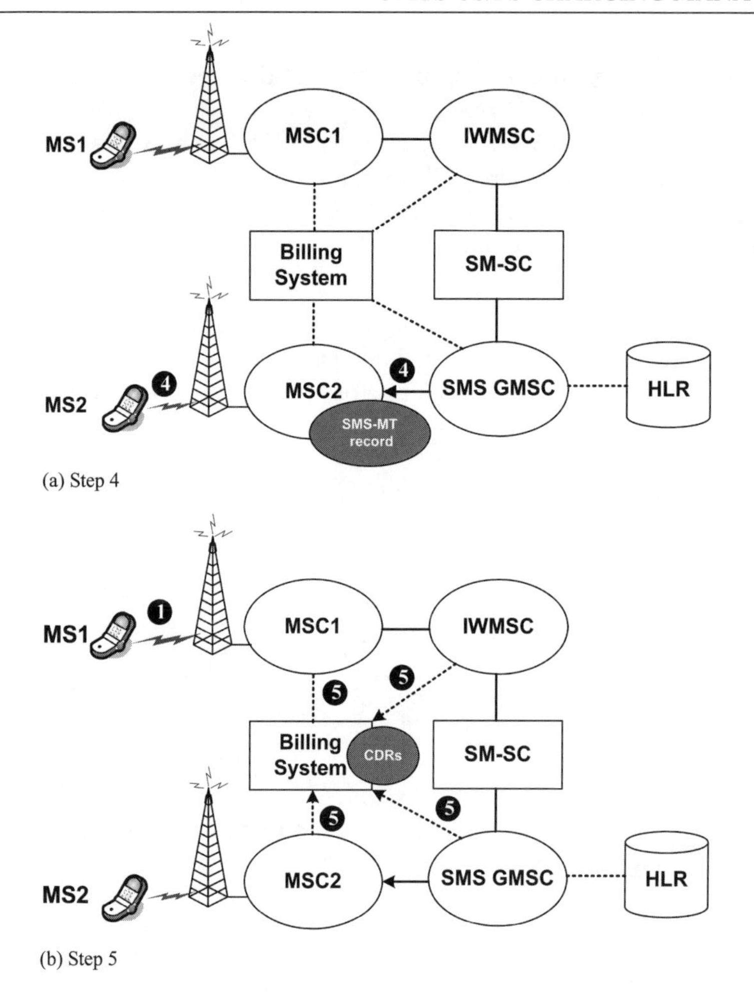

Figure 6.6 An example of SMS delivery (Steps 4 and 5)

Step 1. **[Figure 6.5(a)]** When MS1 sends a short message to MS2, MSC1 creates an *SMS Mobile Originated* (SMS-MO) record and forwards the short message to an MSC called *SMS Inter-Working MSC* (IWMSC).

Step 2. **[Figure 6.5(a)]** The IWMSC creates an SMS-MO interworking record and passes this short message to a *Short Message Service Center* (SM-SC). The SM-SC then forwards the short message to the SMS GMSC of the destination network.

Step 3. **[Figure 6.5(b)]** The GMSC creates an *SMS Mobile Terminated* (SMS-MT) gateway record and queries the HLR to locate the serving MSC of MS2 (MSC2).

Step 4. **[Figure 6.6(a)]** The short message is then delivered to MS2 through MSC2. MSC2 creates an SMS-MT record.

Step 5. **[Figure 6.6(b)]** In this example, the SMS-MO record, the SMS-MO interworking record, the SMS-MT gateway record and the SMS-MT record are generated. When the SMS delivery is finished, these records are forwarded to the billing system.

The SMS-MO record (generated by MSC1 in Figure 6.5(a)) contains the following fields:

- The *Record Type* field indicates "SMS mobile originated".
- The *Served IMSI* field specifies the IMSI of the subscriber (i.e., MS1) sending the short message.
- The *Service Center* field specifies the E.164 number of the SM-SC.
- The *Recording Entity* field specifies the E.164 number of MSC1.
- The *Event Time Stamp* field indicates the time at which the message is received by MSC1.
- The *Message Reference* field indicates a reference number uniquely identifying the short message. This reference number is provided by MS1.

The CDR formats of the SMS-MO interworking record (generated by the IWMSC in Figure 6.5(a)), the SMS-MT gateway record (generated at the SMS GMSC in Figure 6.5(b)) and the SMS-MT record (generated at MSC2 in Figure 6.6(a)) are similar to those for the SMS-MO record, and the details are omitted.

6.2 Packet Switched Service Domain

In the PS service domain, an SGSN (Figure 6.1(d)) plays a role similar to an MSC in the CS service domain. The GGSN (Figure 6.1(e)) interworks the GPRS network to the external *Packet Data Network* (PDN; see Figure 6.1(i)). In a GPRS session, the charging information is collected by the SGSN and the GGSN. These CDRs are sent to the *Charging Gateway* (CG; see Figure 6.1(f)) via the GTP' protocol described in Chapter 4. Then the CG analyzes and possibly consolidates the CDRs from the SGSN/GGSN, and forwards the consolidated data to a billing system. The details are given in the following steps [3GP05c, 3GP05d]:

Step 1. **[Figure 6.7(a)]** Suppose that MS1 accesses a web server in the PDN through GPRS. A *Packet Data Protocol* (PDP) context is created in each of the SGSN and the GGSN when the subscriber activates a GPRS session. In the PDP context activation procedure described in Figure 2.6 of Section 2.3.1, the created PDP context contains a unique *GPRS Charging Identifier* (GCID), charging characteristics (e.g., normal charging, prepaid charging, flat rate charging, or hot billing), the subscriber information (e.g., the IMSI) and the session information (e.g., the PDP type, the PDP address and the negotiated QoS profile). The charging characteristics are supplied by the HLR (Figure 6.1(h)) to

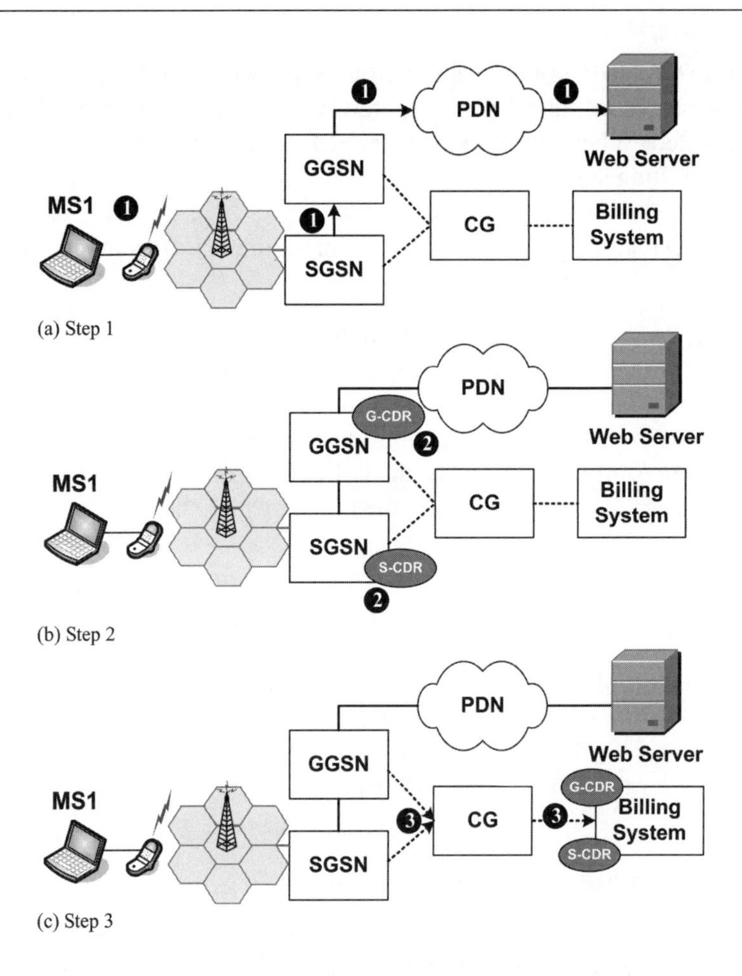

Figure 6.7 An offline charging example of a GPRS session

the SGSN as part of the subscription information. Upon activation of a PDP context, the SGSN forwards the charging characteristics to the GGSN. The GCID is created at the GGSN. The SGSN/GGSN handle the GPRS session according to the charging characteristics methods defined in [3GP05d].

Step 2. **[Figure 6.7(b)]** In the PDP context deactivation described in Figure 2.7 of Section 2.3.1, the SGSN creates the *SGSN generated CDR* (S-CDR) and the GGSN creates the *GGSN generated CDR* (G-CDR) for MS1 to record the served session time and transferred data packets.

Step 3. **[Figure 6.7(c)]** The generated records are subsequently transferred to the CG through the GTP' protocol. The CG transfers the CDRs to the billing system when the GPRS session is complete.

An S-CDR generated in the SGSN (Figure 6.7(b)) contains the following fields [3GP05d]:

- The *Record Type* field indicates "SGSN PDP context record".
- The *Served IMSI* field specifies the identity of the served party (MS1).
- The *Charging ID* field specifies the identity (GCID) of the charging records generated by the SGSNs/GGSN for the same GPRS session.
- The *GGSN Address Used* field contains the control plane IP address of the GGSN. The control plane of GGSN monitors the resources usage of the GPRS session (in the user plane). The user plane IP address of the GGSN is not needed in the S-CDR.
- The *Record Opening Time* field shows the time when the PDP context is activated in this SGSN or when the subsequent partial record is opened.
- The *Duration* field indicates the recording period of this session in the SGSN.
- The *Cause for Record Closing* field describes the reason for the closure of the S-CDR, such as "data volume limit", "time limit" for partial record generation, "PDP context release", and "GPRS detach" for normal release.
- The *Charging Characteristics* field indicates the charging mechanism accountable for the GPRS connection.

A G-CDR (generated by the GGSN in Figure 6.7(b)) contains the following fields:

- The *Record Type* field specifies "GGSN PDP context record".
- The *SGSN Address Used* field lists the SGSN addresses used in this GPRS connection. This field contains more than one SGSN address if the MS1 moves across several SGSNs during the GPRS session.
- The *Record Opening Time* field shows the time when the PDP context is activated or when the subsequent partial record is opened in the GGSN.
- The *Duration* field indicates the recording period of the G-CDR.
- The *Served IMSI*, the *Charging ID*, the *Cause for Record Closing*, the *Charging Characteristics* fields are similar to those in the S-CDR.

In GPRS, a long session may be covered by several partial CDRs. A partial CDR provides information for a portion of a GPRS session. Partial closure of the CDR is triggered by the SGSN/GGSN in one of the following three cases:

- The session time (duration) limit or the data volume limit specified in the charging characteristics profile is reached.
- The MS moves from one SGSN to another SGSN during the GPRS session.
- The maximum number of changes in charging conditions (e.g., due to QoS changes or tariff time changes) is reached.

Note that all partial CDRs are identified by the same GCID. The GCID and the GGSN address uniquely identify a GPRS session. In an activated GPRS session, the GGSN address is always the same while the SGSN address may be changed due to the MS mobility. During an MS movement, the complete PDP context is forwarded from the old SGSN to the new SGSN by exercising the update PDP context procedure defined in [3GP05c]. Therefore, the resulting charging record generated in the new SGSN uses the same GCID specified in the PDP context. When the GPRS session is complete, these partial records are correlated in the CG and the billing system based on the GCID and the GGSN address.

6.3 Concluding Remarks

This chapter described the UMTS Release 5 (R5) charging functionalities that allow the UMTS network nodes to generate the CDRs accurately. We introduced the charging management on the circuit switched service domain and the packet switched service domain. Specifically, we provided examples to illustrate when the charging records are generated, and what kinds of information are contained in these records.

An overview on R5 charging management can be found in [3GP05a]. More details of charging for the circuit switched domain and the packet switched domain can be found in [3GP05b] and [3GP05d], respectively. In the next chapter, we will elaborate on the UMTS R5 charging for IMS and MMS application service domains.

Review Questions

1. Describe the GSM network components that are involved in a CS offline charging call and a CS online charging call.
2. Can gprsSSF be implemented in the GGSN? Why or why not?
3. Does gsmSCF interact with the billing system? Why or why not?
4. Describe CS call setup from a mobile station to another mobile station, and specify the CDRs created during call setup.
5. Give an example when the HLR interrogation record is used by a mobile operator.
6. How does a mobile operator measure the call holding time statistics?
7. Describe the differences between the MOC and the MTC records.
8. Is it required to record the called number in an MOC record? Why or why not?
9. For the prepaid call example in the CS service domain, will there be any CDRs created in the PSTN?
10. Suppose that a call was set up between Phone1 and MS2 through the path (a)–(b)–(c) in Figure 6.8. During the call session, MS2 moved from MSC1 to MSC2 (and the call path becomes (a)–(b)–(d)–(e)). Then MS2 moved from MSC2 to MSC3 again (and the call path becomes (a)–(b)–(d)–(f)–(g)). What is the life time of the CDR

in MSC1? Are CDRs created in MSC2 and MSC3? If so, how are these CDRs consolidated?

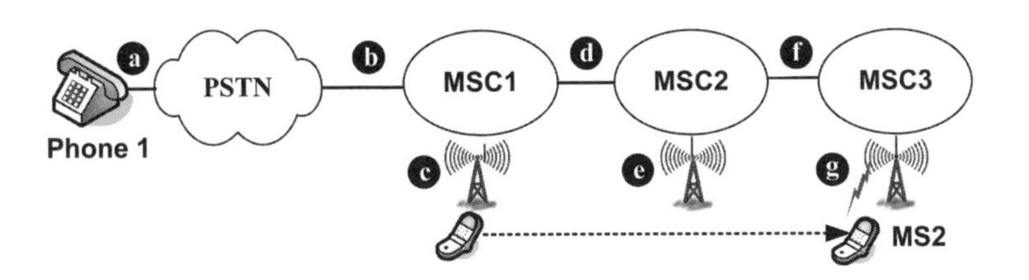

Figure 6.8 MS movement during a call

11. What are the differences between the IWMSC in Figure 6.5 and the outgoing GMSC in Figure 6.2?
12. Compare the SMS-MO record with the MOC record. What are the major differences?
13. For offline CDR creation, what is the major difference between GSM and GPRS?
14. If GGSN is changed in a GPRS session, can we still use the same GCID for the session?
15. Should we record the user plane IP address of the GGSN in the S-CDR record? Why or why not?
16. Which fields in the S-CDR are not found in the G-CDR. Why?
17. Describe the online charging example in the PS service domain.

References

[3GP02] 3GPP, 3rd Generation Partnership Project; Technical Specification Group Services and Systems Aspects; Circuit Teleservices supported by a Public Land Mobile Network (PLMN) (Release 5), 3G TS 22.003 version 5.2.0 (2002-06), 2002.

[3GP04] 3GPP, 3rd Generation Partnership Project; Technical Specification Group Core Network; Customized Applications for Mobile network Enhanced Logic (CAMEL) Phase 4; CAMEL Application Part (CAP) specification (Release 5), 3G TS 29.078 version 5.9.0 (2004-09), 2004.

[3GP05a] 3GPP, 3rd Generation Partnership Project; Technical Specification Group Services and Systems Aspects; Telecommunication management; Charging management; Charging principles (Release 5), 3G TS 32.200 version 5.9.0 (2005-09), 2005.

[3GP05b] 3GPP, 3rd Generation Partnership Project; Technical Specification Group Services and Systems Aspects; Telecommunication management; Charging management; Charging data description for the Circuit Switched (CS) domain (Release 5), 3G TS 32.205 version 5.9.0 (2005-03), 2005.

[3GP05c] 3GPP, 3rd Generation Partnership Project; Technical Specification Group Services and Systems Aspects; General Packet Radio Service (GPRS); Service description; Stage 2 (Release 5), 3G TS 23.060 version 5.13.0 (2006-12), 2006.

[3GP05d] 3GPP, 3rd Generation Partnership Project; Technical Specification Group Services and Systems Aspects; Telecommunication Management; Charging Management; Charging data description for the Packet Switched (PS) domain (Release 5), 3G TS 32.215 version 5.9.0 (2005-06), 2005.

[3GP05e] 3GPP, 3rd Generation Partnership Project; Technical Specification Group Services and System Aspects; Customized Applications for Mobile network Enhanced Logic (CAMEL); Service description; Stage 1 (Release 5), 3G TS 22.078 version 5.15.0 (2005-03), 2005.

[3GP06] 3GPP, 3rd Generation Partnership Project; Technical Specification Group Core Network; Customized Applications for Mobile network Enhanced Logic (CAMEL) Phase 4; Stage 2 (Release 5), 3G TS 23.078 version 5.11.0 (2006-06), 2006.

[Lin01] Lin, Y.-B. and Chlamtac, I., *Wireless and Mobile Network Architectures*. John Wiley & Sons, Ltd., Chichester, UK, 2001.

7

IMS and MMS Offline Charging Management

Charging of the *IP Multimedia core network Subsystem* (IMS) aims to provide a single, cost-effective software solution for telecom operators to consolidate charging capabilities while supporting all aspects of the IMS standard. Hopefully, the telecom operators can launch new services, pricing plans and loyalty programs more efficiently and cost-effectively than ever before. In Chapter 6 we described how UMTS Release 5 (R5) defines online and offline charging for the CS and the PS service domains. This chapter shows how R5 supports the charging functionalities for IMS and MMS offline charging. The IMS adds Diameter-based interfaces and applications to create fully-compliant elements of a charging solution [3GP02]. To interact with the IMS call session control function for offline charging, the 3GPP standard defines the *Charging Collection Function* (CCF). This function constructs, correlates, formats, and transfers the information about charging events to the billing system.

In this chapter, two charging scenarios are given to explain how the CDRs are generated in offline IMS and MMS service domains. Then we introduce the mediation function that mediates the charging data sent from a network node to the billing system. Note that the interaction between the *Online Charging System* (OCS) and the billing system was not defined in this release. A complete OCS architecture defined in UMTS Release 6 will be elaborated in Chapter 8.

7.1 Offline Charging for IMS

This section describes the UMTS R5 resource control and the charging architecture for offline IMS services (see Figure 7.1). A major component in this architecture is the CCF (Figure 7.1(a)). The IMS nodes involved in offline charging include MRF (Figure 7.1(c)), MGCF (Figure 7.1(d)), S-CSCF (Figure 7.1(e)), P-CSCF

Charging for Mobile All-IP Telecommunications Yi-Bing Lin and Sok-Ian Sou

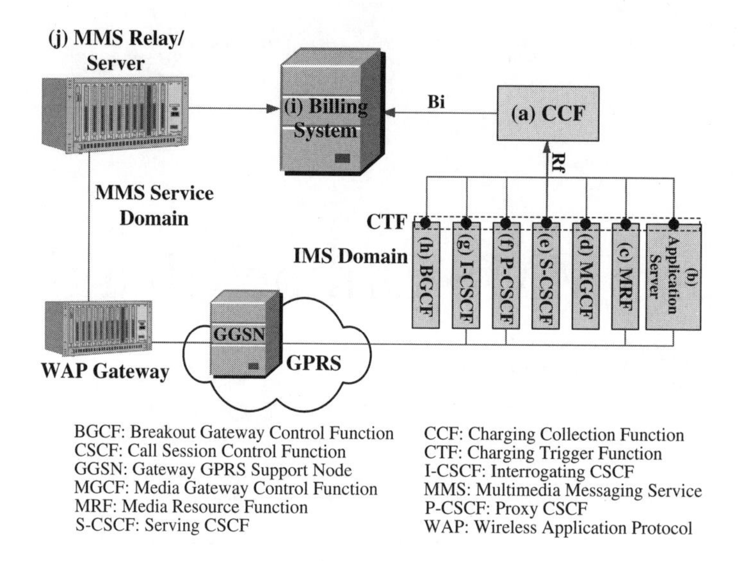

Figure 7.1 The IMS and MMS service domains in UMTS R5 offline charging (the bullets in the IMS domain represent the Charging Trigger Functions (CTFs))

(Figure 7.1 (f)), I-CSCF (Figure 7.1(g)) and BGCF (Figure 7.1(h)). In this architecture, the application server (Figure 7.1(b)) offering value-added IP multimedia service resides either in the user's home network or in a third-party location [3GP06b]. The offline *Charging Trigger Function* (CTF; represented by the bullets in Figure 7.1) is a mandatory function integrated in each IMS network node to provide offline charging functionality. The CTF is implemented in an IMS/GPRS network node, which is responsible for monitoring service usage and generating charging events based on it. In other words, the CTF provides metrics that identify the users and their consumption of network resources, and makes chargeable events from these metrics. It then sends the offline charging information to the CCF. The CCF forwards the processed CDRs to the billing system (Figure 7.1(i)). This model utilizes the Diameter protocol described in Chapter 5, where the CCF is a Diameter server and the IMS nodes are Diameter clients.

An IMS node uses the Diameter Accounting Request (ACR) to send the offline charging information to the CCF. The CCF constructs and formats the CDRs, and then acknowledges the IMS node with the Accounting Answer (ACA) message. The CCF opens an IMS session CDR when it receives a Diameter ACR message with *Accounting-Record-Type* "start". During the IMS session, the accounting information can be updated when the IMS node sends the ACA message with *Accounting-Record-Type* "interim" to the CCF. The CCF closes the session CDR upon receipt of an ACR message with *Accounting-Record-Type* "stop". For event-based procedures such as user registration or HSS interrogation, accounting information is transferred from an IMS node to the CCF using the ACR message with *Accounting-Record-Type* "event". The CCF can be implemented as a network node or as a functionality residing in an IMS node.

Figures 7.2 and 7.3 depict an offline charging example. The IMS call procedure is given in the following steps:

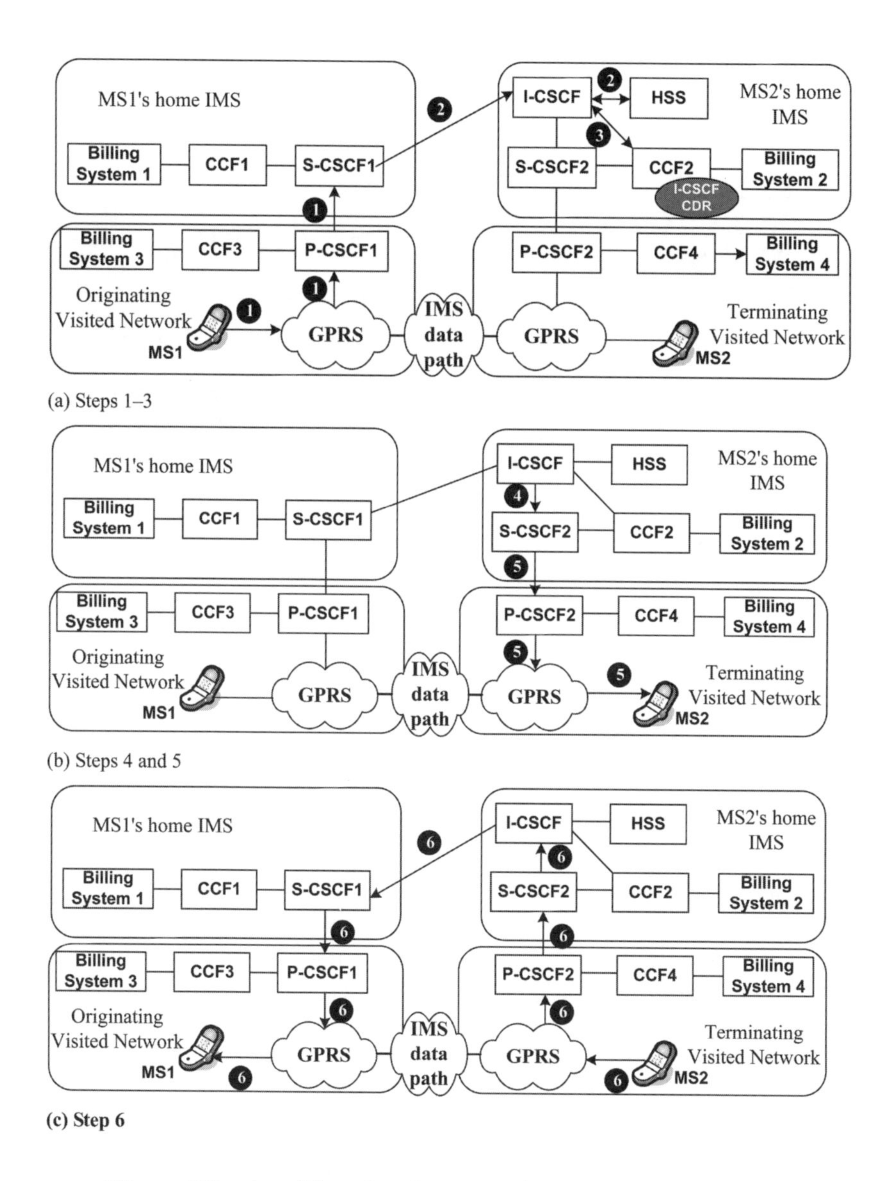

Figure 7.2 An offline charging example of IMS call (Steps 1–6)

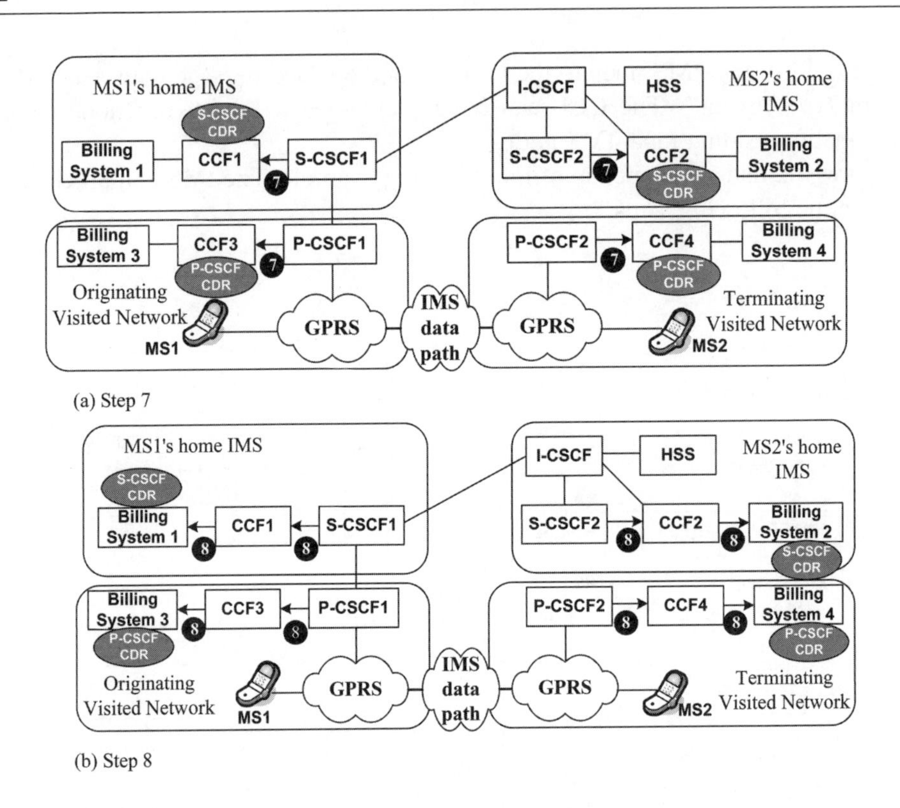

Figure 7.3 An offline charging example of IMS call (Steps 7 and 8)

Step 1. **[Figure 7.2(a)]** MS1 makes an IMS call to MS2 by sending a SIP INVITE request [3GP06b]. The *Session Description Protocol* (SDP) contained in this message specifies related media parameters (such as codec, media type, and bandwidth). The INVITE message is first forwarded to the proxy CSCF (i.e., P-CSCF1) and then to the serving CSCF (i.e., S-CSCF1) of MS1.

Step 2. **[Figure 7.2(a)]** S-CSCF1 sends the INVITE message to the interrogating CSCF of MS2 (i.e., I-CSCF). The I-CSCF interrogates the HSS to identify the serving CSCF for MS2 (i.e., S-CSCF2).

Step 3. **[Figure 7.2(a)]** After interrogating the HSS, the I-CSCF sends the ACR message with *Accounting-Record-Type* "event" to CCF2. Based on the event type, CCF2 creates an I-CSCF CDR. Then CCF2 replies with the ACA message to the I-CSCF.

Step 4. **[Figure 7.2(b)]** The I-CSCF sends the INVITE request to S-CSCF2.

Step 5. **[Figure 7.2(b)]** Through S-CSCF2 and the proxy CSCF of MS2 (i.e., P-CSCF2), the INVITE request is passed to MS2.

Step 6. **[Figure 7.2(c)]** If MS2 accepts the call, the SIP 200 OK response is sent back to MS1.

In the above steps, when the P-CSCFs (P-CSCF1 and P-CSCF2) receive the SIP messages from the MSs (MS1 and MS2), they send the SDP parameters to the *Policy Decision Function* (PDF). The PDF then authorizes the related media parameters according to the users' media requirements and the local policy. Details of the PDF will be elaborated in Section 9.4. After authorization, the authorized media parameters are returned to the MS and the resources for setting up the transmission bearer are reserved. The PDF forwards the related IP QoS control parameter to the GGSN. To execute the QoS control policy, the GGSN analyzes the source and destination IP addresses. Then it controls and filters the IP flow following the UMTS R6 and R7 specifications to be described in Chapter 9.

Step 7. **[Figure 7.3(a)]** Upon receipt of the 200 OK response, P-CSCF2 sends an ACR message to CCF4, S-CSCF2 sends an ACR message to CCF2, S-CSCF1 sends an ACR message to CCF1, and P-CSCF1 sends an ACR message to CCF3. These ACR messages include the *Accounting-Record-Type* "start" to indicate that the CDRs are opened for an IMS call session. The CCFs create the related CDRs (i.e., S-CSCF CDRs and P-CSCF CDRs) and reply with the ACA messages.

Step 8. **[Figure 7.3(b)]** When the call is terminated, P-CSCF1, S-CSCF1, P-CSCF2 and S-CSCF2 send the ACR messages with *Accounting-Record-Type* "stop" to the corresponding CCFs for closing the CDRs. These CDRs are then transferred to the corresponding billing systems for offline processing.

A CSCF CDR (created in Figure 7.2(a) and Figure 7.3(a)) contains the following fields [3GP06a]:

- The *Record Type* field specifies the CSCF type (e.g., P-CSCF CDR or S-CSCF CDR).
- The *Role of Node* field indicates the role of the CSCF (either originating or terminating) [3GP06b]. The originating nodes are the CSCFs (i.e., P-CSCF1 and S-CSCF1 in Figure 7.2) of the calling party (i.e., MS1). The terminating nodes are the CSCFs (i.e., S-CSCF2 and P-CSCF2 in Figure 7.2) of the called party (i.e., MS2).
- The *Node Address* field specifies the node providing the accounting information. This address may be an IP address or a *Fully Qualified Domain Name* (FQDN) of the IMS node (i.e., the FQDN for P-CSCF1, P-CSCF-2, S-CSCF1 or S-CSCF2 in Figure 7.2).
- The *Session ID* field contains the call ID specified in the SIP message.
- The *Calling* and the *Called Party Address* fields specify the addresses (e.g., SIP *Uniform Resource Locator* (URL) or TEL URL) of the calling party (MS1) and the called party (MS2), respectively.

- The *Served Party IP Address* field contains the IP address of either the calling or the called party.
- The *Service Request Time Stamp* field indicates the time at which the service is requested.
- The *Service Delivery Start/End Time Stamp* field indicates the time when the IMS session is started/ released.
- The *Record Opening/Closure Time* field indicates the time when the CCF opens/closes the CDR record.
- The *Local Record Sequence Number* field includes a unique record number created by the IMS node.
- The (optional) *Record Sequence Number* field contains a running sequence number employed to link the partial records generated by the CCF.
- The *Inter Operator Identifiers* (IOIs) field holds the globally unique identifiers of home IMS networks for the calling and the called parties, respectively. More details about the IOI will be given in Section 7.2.
- The *Cause for Record Closing* field contains the reason for a CDR release. The reason can be "end of session", "service change", "CCF initiated record closure", and so on.
- The *IMS Charging Identifier* (ICID) field is a unique identifier generated by the IMS for the SIP session.
- The *List of SDP Media Components* field contains the SIP request timestamp, SIP response timestamp, SDP media components (i.e., SDP media name, SDP media description, GPRS charging ID (i.e., GCID)) and media initiator flag.
- The *GGSN Address* field specifies the control plane IP address of the GGSN.

In an IMS session, it is possible that the CCF generates multiple partial records due to changes of session characteristics (or changes of charging conditions). These events include tariff switching, QoS changed, or time (volume) limit exceeded. In this case, the *Record Sequence Number* field is used to link the partial records generated by the CCF.

7.2 IMS Charging Correlation

This section describes how IMS charging is integrated with the PS service domain charging described in Section 6.2. Several charging correlations exist in an IMS session, including:

- correlation between CDRs generated by different IMS nodes;
- correlation between CDRs generated by the GPRS support nodes and CDRs generated by IMS nodes; and
- correlation between CDRs generated by different operators.

These charging correlations are described as follows:

The CDRs generated in the IMS nodes are correlated by an *IMS Charging Identifier* (ICID). This identifier is globally unique across all 3GPP IMS networks for a period of at least one month. Expiration of an ICID can be detected by using IMS node-specific information, e.g., high-granularity time information and location information. A new ICID is generated for an IMS session at the first IMS node that processes the SIP INVITE message. The ICID is included in all subsequent SIP messages (e.g., 200 OK, UPDATE, and BYE) for that IMS session until the session is terminated. For the example shown in Figure 7.2, an ICID is generated by P-CSCF1 for mobile-originated calls. The SIP request includes the ICID specified in the P-Charging-Vector header (see Section 5.6.2). This ICID is passed along the SIP signaling path to all involved IMS nodes. Each IMS node that processes the SIP request retrieves the ICID and includes it in the CDRs generated later. In a CSCF CDR, the ICID is listed in the *IMS Charging Identifier* field. Therefore, CDRs from different IMS nodes within an IMS session can be correlated in the billing system through the ICID.

In an IMS session, the media (e.g., voice or video) packets are transferred through the GPRS bearer session. The GPRS charging information (e.g., the GGSN address and the GCID) can be used to correlate the GPRS CDRs with the IMS CDRs. The GPRS charging information for a media session is included in the P-Charging-Vector header, and is passed from the GGSN to the P-CSCF (e.g., P-CSCF1 and P-CSCF2 in Figure 7.2). Note that the GPRS charging information for the originating network is used only within the originating network. Similarly, the GPRS charging information for the terminating network is used only within the terminating network. In the SIP signaling path, the GGSN address, the GCID and the ICID are passed from the P-CSCF to the S-CSCF and other IMS nodes. The S-CSCF also passes this information to the CCF. In a CSCF CDR, the related GCID is included in the *List of SDP Media Components* field.

As described in Section 5.6.2, charging correlation among different operators is achieved through the *Inter Operator Identifier* (IOI), a globally unique identifier shared between operator networks and service/content providers [3GP08]. With IOIs, the home IMS networks for both call parties can settle the account with each other. The originating IOI and the terminating IOI are exchanged by the SIP request and response messages. The originating S-CSCF (e.g., S-CSCF1 in Figure 7.2) includes the originating IOI in the P-Charging-Vector header of the SIP request. In a CSCF CDR, the IOIs are listed in the *Inter Operator Identifiers* field. Upon receipt of the SIP message, the terminating S-CSCF (e.g., S-CSCF2 in Figure 7.2) retrieves the originating IOI of calling party's home IMS network. It then includes the terminating IOI in the P-Charging-Vector header of the SIP response message. Through the SIP response message, the originating S-CSCF retrieves the terminating IOI of the called party's home IMS network.

IMS also supports online charging capabilities through the OCS, where an IMS node or an application server interacts with the OCS in real time to access the user accounts and controls the charges related to service usage. Details of the OCS and the online charging examples will be given in Chapter 8.

7.3 Multimedia Messaging Service Domain

In the R5 *Multimedia Messaging Service* (MMS) domain, CDRs generated from the MMS Relay/Server (Figure 7.1(j)) are transferred directly to the billing system (Figure 7.1(i)) for offline charging. MMS delivers multimedia objects including texts, images, audio and video. In the MMS architecture illustrated in Figure 2.9 (Section 2.3.3), the MMS charging records are generated by the MMS Relay/Server when it sends multimedia messages to the MMS *User Agent* (UA) or receives multimedia messages from the MMS UA [3GP04]. Several CDRs are generated in an MMS delivery. Figures 7.4 and 7.5 depict the MMS delivery procedure in the following steps [3GP05]:

Step 1. **[Figure 7.4(a)]** An MMS user agent (UA1) sends a multimedia message to another MMS user agent (UA2). The message (encapsulated in a WAP POST message) is first sent to the MMS Relay/Server. An *Originator MM1 Submission Record* (O1S-CDR) is created in the MMS Relay/Server.

Step 2. **[Figure 7.4(a)]** The MMS Relay/Server sends a notification message (encapsulated in a WAP PUSH message) to UA2.

Step 3. **[Figure 7.4(b)]** After receiving the notification from the MMS Relay/Server, UA2 requests to receive the multimedia message through a WAP GET message.

Step 4. **[Figure 7.4(b)]** The multimedia message is sent to UA2 through the WAP GET response at Step 4(a). UA2 retrieves the message and acknowledges the MMS Relay/Server. A *Recipient MM1 Retrieve Record* (R1Rt-CDR) and a *Recipient MM1 Acknowledgment Record* (R1A-CDR) are created in the MMS Relay/Server at Step 4(b).

Step 5. **[Figure 7.5(a)]** Through the WAP PUSH message, the MMS Relay/Server informs UA1 that the multimedia message is delivered successfully. An *Originator MM1 Delivery Report Record* (O1D-CDR) is created at the MMS Relay/Server.

Step 6. **[Figure 7.5(b)]** After UA2 has received the multimedia message, it sends a WAP PUSH message to the MMS Relay/Server to indicate that the message is read. Then a *Recipient MM1 Read-reply Recipient Record* (R1RR-CDR) is created in the MMS Relay/Server.

Step 7. **[Figure 7.5(b)]** The CDRs generated in the previous steps are subsequently transferred to the billing system after the MMS message is delivered.

In the above example, the MMS Relay/Server plays both roles as the originator and recipient MMS Relay/Server. If UA1 and UA2 are located in different UMTS

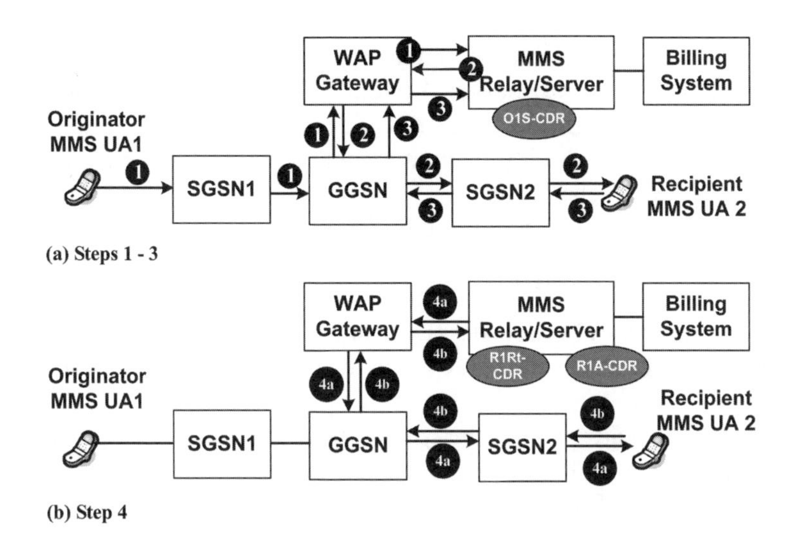

Figure 7.4 Offline charging for MMS delivery (Steps 1–4)

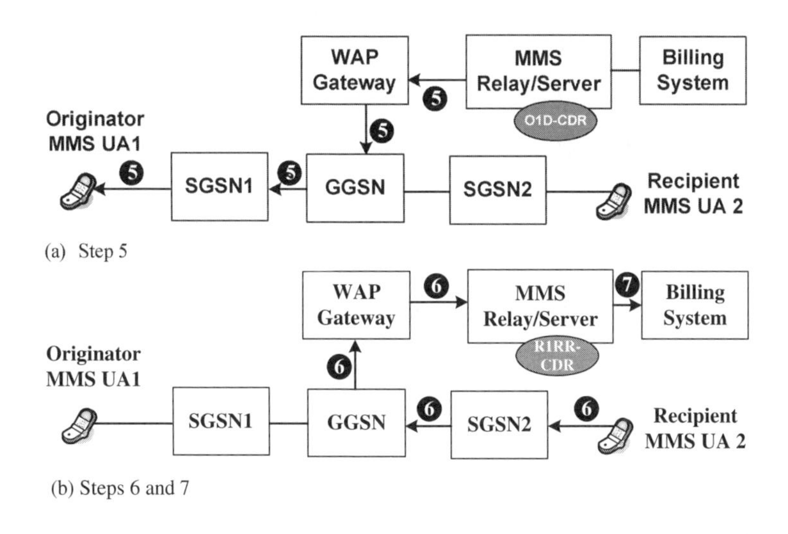

Figure 7.5 Offline charging for MMS delivery (Steps 5–7)

networks, then the MMS delivery will involve an originator MMS Relay/Server (for UA1) and a recipient MMS Relay/Server (for UA2). In this case, the O1S-CDR and the O1D-CDR are created in the originator MMS Relay/Server, and the R1Rt-CDR, R1A-CDR and R1RR-CDR are created in the recipient MMS Relay/Server.

An O1S-CDR generated by the originator MMS Relay/Server (at Step 1 in Figure 7.4(a)) contains the following fields [3GP05]:

- The *Record Type* field indicates "Originator MM1 submission record".
- The *Originator MMS Relay/Server Address* field specifies the IP address or the domain name of the originator MMS Relay/Server.
- The *Message ID* field specifies the multimedia message identification provided by the originator MMS Relay/Server.
- The *Recipient Address List* field specifies the addresses (e.g., email addresses or MSISDNs) for the recipients of the multimedia message (e.g., UA2 in Figure 7.4).
- The *Access Correlation* field indicates a unique identifier (e.g., the GCID) delivered by the access network of the sender (UA1 in Figure 7.4). This information may be used for correlation of the MMS CDR with the GPRS CDR in the PS service domain.
- The *Content Type* field specifies the type of the multimedia message content, such as plain text, speech, audio, image, video, and so on.
- The *MM Component List* field specifies the media components including the size of each component.
- The *Message Size* field indicates the size of the multimedia message content.
- The *Record Time Stamp* field specifies the time of CDR generation.
- The *Local Record Sequence Number* field includes a record number created by the originator MMS Relay/Server.

An R1Rt-CDR generated by the recipient MMS Relay/Server (at Step 4 in Figure 7.4(b)) contains the following fields:

- The *Record Type* field indicates "Recipient MM1 retrieve record".
- The *Recipient MMS Relay/Server Address* field specifies the IP address (or the domain name) of the recipient MMS Relay/Server.
- The *Recipient Address* field specifies the address (e.g., an email address or an MSISDN) of the multimedia message recipient.
- The *Message Reference* field contains the location (e.g., a *Uniform Resource Identifier* (URI)) that can be used for retrieving the multimedia message from the recipient MMS Relay/Server.
- The *Original MM Content* field specifies a set of parameters including the content type, the message size and the media component list of the original multimedia message.
- The *Submission Time* field specifies the time at which the multimedia message was submitted or forwarded.
- The *MMS Status Code* and the *Status Text* fields indicate the status code (e.g., retrieved, rejected, etc.) and the detailed description for the status.
- The *Message ID*, the *Access Correlation*, the *Record Time Stamp*, and the *Local Record Sequence Number* fields are similar to those in the O1S-CDR.

An R1A-CDR generated by the recipient MMS Relay/Server (at Step 4 in Figure 7.4(b)) contains the following fields:

- The *Record Type* field indicates "Recipient MM1 acknowledgment record".
- The *Recipient MMS Relay/Server Address*, the *Recipient Address*, the *Message ID*, the *Access Correlation*, the *MMS Status Code*, the *Status Text*, the *Record Time Stamp*, and the *Local Record Sequence Number* fields are similar to those in the R1Rt-CDR.

An O1D-CDR generated by the originator MMS Relay/Server (at Step 5 in Figure 7.5(a)) contains the following fields:

- The *Record Type* field indicates "Originator MM1 delivery report record".
- The *Originator MMS Relay/Server Address* and the *Recipient MMS Relay/Server Address* fields specify the IP addresses or the domain names of the originator and recipient MMS Relays/Servers.
- The *Originator Address* and the *Recipient Address* fields include the addresses (e.g., email addresses or MSISDNs) of the sender and the receiver of the multimedia message.
- The *Message ID*, the *Access Correlation*, the *Record Time Stamp*, and the *Local Record Sequence Number* fields are similar to those in the O1S-CDR.

An R1RR-CDR generated by the recipient MMS Relay/Server (at Step 6 in Figure 7.5(b)) contains the following fields:

- The *Record Type* field indicates "Recipient MM1 Read reply recipient record".
- The *Recipient MMS Relay/Server Address* field specifies the IP address or the domain name of the recipient MMS Relay/Server.
- The *Originator Address* and the *Recipient Address* fields specify the addresses (e.g., email addresses or MSISDNs) of the sender and the receiver of the multimedia message.
- The *Message ID*, the *Access Correlation*, the *MMS Status Code*, the *Status Text*, the *Record Time Stamp*, and the *Local Record Sequence Number* fields are similar to those in the O1S-CDR.

When UA1 sends a multicast multimedia message, the recipient address list is included in the O1S-CDR. These recipients may be served by different access networks (e.g., 2G or 3G), and each R1RR-CDR is generated when the MMS is read by each recipient. Therefore, a number of recipient CDRs (e.g., R1Rt-CDRs and R1RR-CDRs) may be generated for a multicast MMS delivery. These CDRs should be collected and settled with the O1S-CDR. We note that in Taiwan, only the sender of MMS delivery will be charged, and R1Rt-CDR/R1RR-CDRs are not used for charging.

As a final remark, if the originator and the recipient MMS Relays/Servers locate in different networks, several extra CDRs will be generated to log the events between the MMS interfaces such as MM1 and MM4 shown in Figure 7.6. In this figure, MM1 is the interface between the MMS UA and the MMS Relay/Server; MM4 is the interface between the MMS Relay/Server and another MMS Relay/Server [3GP04]. More details for these CDRs can be found in [3GP05] (see also review question 10 at the end of this chapter).

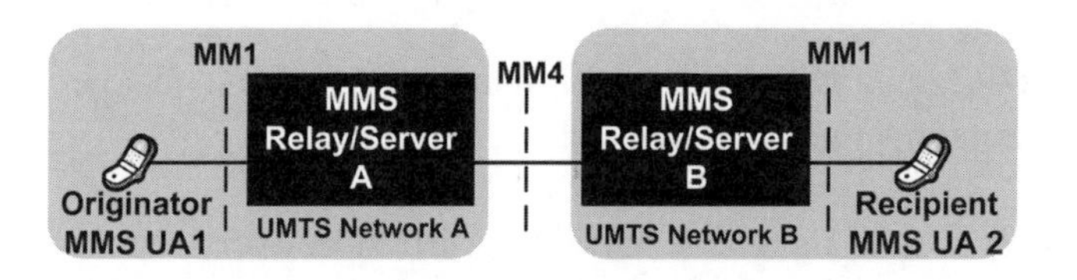

Figure 7.6 Network interworking for MMS delivery

7.4 Mediation Device

In the telecommunications industry, a billing system that integrates customer care functionalities is usually called a *Customer Care & Billing System* (CCBS). A CCBS must be able to process huge amounts of charging records from a large number of network nodes. The above statement is particularly true for the IMS and the MMS service domains (for example, multiple CDRs are generated for one MMS delivery as described in the previous section). These records may have different structures depending on the equipment vendors. Since these structures are typically modified when new releases or new equipment are introduced, both performance and maintainability of a CCBS are critical issues. In order to reduce the maintenance costs and improve the performance of the CCBS, a telecom operator often uses a billing *Mediation Device* (MD; Figure 7.7(b)) to collect CDRs from different network nodes such as MSC and GGSN (Figure 7.7(c)). An MD may be implemented as a standalone

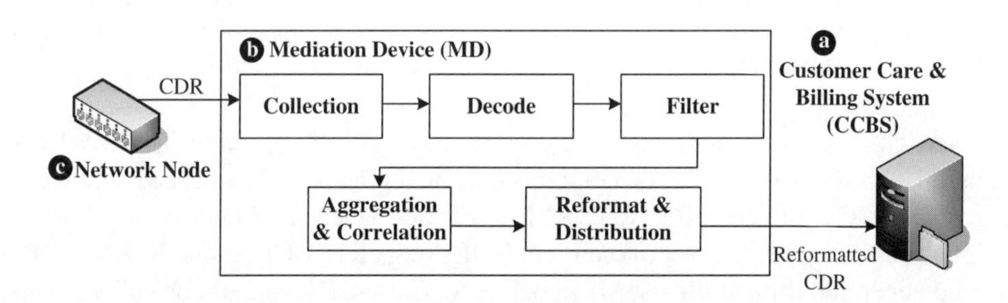

Figure 7.7 Mediation device for billing system

device or as a component of the CCBS (Figure 7.7(a)). For example, in Figure 7.1, an MD can be a standalone device located between the CCF (the MMS server) and the billing system (i.e., a CCBS).

By filtering out unnecessary information and consolidating charging data into a suitable format for processing, the MD reduces the number of billing records sent to the billing system. The data processed by the MD are either continuously or periodically sent to the CCBS. The functionalities of the MD are described as follows:

- Collection: This function collects CDRs generated from different network nodes. The CDRs are collected in either push or pull modes via FTP.
- Decode: This function interprets the received CDRs with various formats (e.g., ASN.1 or XML), and fills them into the standard data fields.
- Filter: This function filters out unnecessary CDRs.
- Aggregation & Correlation: This function consolidates several related CDRs into a single record.
- Reformat & Distribution: This function transforms the CDRs into a suitable formatted CDR for processing in the billing system. These reformatted CDRs contain the data fields that are needed in the billing system. The CDRs are then sent to the billing system for statistics, settlement, or accounting purposes.

7.5 Concluding Remarks

This chapter introduced the UMTS offline charging functionalities for the IMS network and the MMS application. Specifically, we provided examples to illustrate when the charging records are generated, and what kinds of information are contained in these records. We also described the mediation device functions that consolidate the charging data generated by the network nodes before they are sent to the billing system.

An overview on UMTS Release 5 charging management can be found in [3GP05]. The details for IMS charging functionality can be found in [3GP06a, 3GP06b], and will be elaborated more in Chapter 8. Details for MMS and its charging functionality can be found in [3GP04, 3GP05].

Review Questions

1. Describe the IMS call setup procedure. How many CDRs will be created?
2. Describe the charging correlation between the GPRS PS domain and the IMS.

3. Describe ICID, GCID and IOI. How do these identities relate to each other?
4. Why is the ICID globally unique across all 3GPP IMS networks for a period of at least one month? How can global unique be guaranteed?
5. How is WAP utilized in the MMS application service?
6. Describe the CDRs generated during an MMS delivery.
7. In IMS, an ICID is defined. In MMS, no specific CID is defined. Why?
8. When will an email address be filled in the Recipient Address field of the R1Rt-CDR?
9. Describe the difference in the "Recipient Address" specified in O1S-CDR and R1Rt-CDR.
10. Draw the message flow for MMS delivery that involves two MMS Relays/Servers depicted in Figure 7.8.

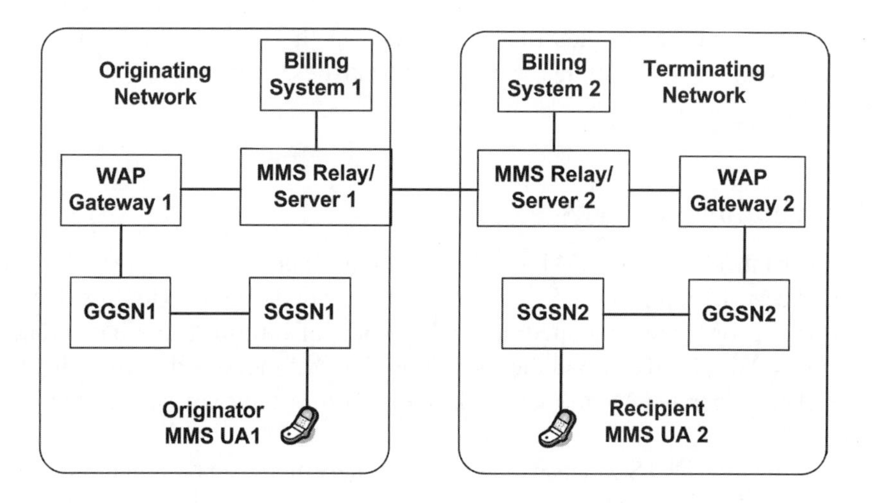

Figure 7.8 MMS delivery involving two MMS relays/servers

11. Describe the Mediation Device (MD) functionalities. Should a telecom operator install an independent MD node or integrate the MD functions as a part of the billing system? What are the tradeoffs of these two approaches?

References

[3GP02] 3GPP, 3rd Generation Partnership Project; Technical Specification Group Services and Systems Aspects; Charging implications of IMS architecture (Release 5), 3G TR 23.815 version 5.0.0 (2002-03), 2002.

[3GP04] 3GPP, 3rd Generation Partnership Project; Technical Specification Group Terminals; Multimedia Messaging Service (MMS); Functional description; Stage 2 (Release 5), 3G TS 23.140 version 5.11.0 (2004-06), 2004.

[3GP05] 3GPP, 3rd Generation Partnership Project; Technical Specification Group; Telecommunication management; Charging management; Charging data description for application services (Release 5), 3G TS 32.235 version 5.5.0 (2005-09), 2005.

[3GP06a] 3GPP, 3rd Generation Partnership Project; Technical Specification Group Services and Systems Aspects; Telecommunication Management; Charging Management; Charging data description for the IP Multimedia Subsystem (IMS) (Release 5), 3G TS 32.225 version 5.11.0 (2006-03), 2006.

[3GP06b] 3GPP, 3rd Generation Partnership Project; Technical Specification Group Core Network; IP Multimedia Subsystem (IMS); Stage 2 (Release 5), 3G TS 23.228 version 5.15.0 (2006-06), 2006.

[3GP08] 3GPP, 3rd Generation Partnership Project; Technical Specification Group Core Network and Terminals; IP Multimedia Call Control Protocol based on Session Initiation Protocol (SIP) and Session Description Protocol (SDP); Stage 3 (Release 6), 3G TS 24.229 version 6.18.0 (2008-06), 2008.

8

UMTS Online Charging

This chapter describes the UMTS Release 6 (R6) charging system that allows the UMTS network nodes to generate and deliver the online CDRs. In Chapter 9, we will show how UMTS R7 charging enhances the R6 version by introducing *Policy and Charging Control* (PCC) [3GP07b]. Figure 8.1 illustrates the logical architecture of online charging. In this architecture, UMTS R6 introduces the convergent charging solution focusing on the *Online Charging System* (OCS; Figure 8.1(a)), the *Traffic Plane Function* (TPF; Figure 8.1(b)) and the *Charging Rules Function* (CRF; Figure 8.1(c)). Details of the TPF and the CRF will be described in Chapter 9. This chapter describes the functionalities of the OCS components. Then we use IMS services as examples to depict how the OCS interacts with the UMTS/IMS network [3GP06a, 3GP06b, 3GP07c].

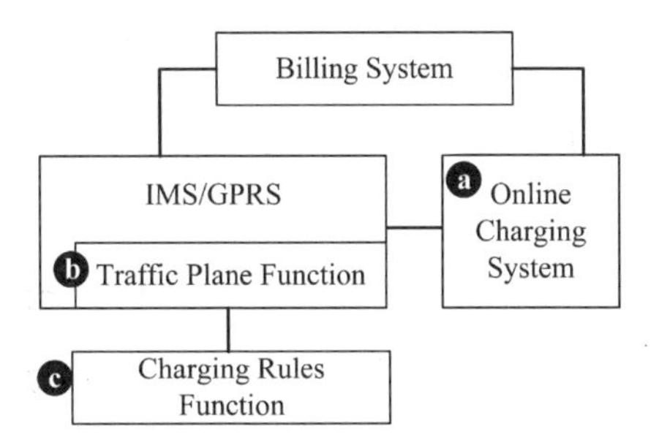

Figure 8.1 Components involved in UMTS online charging

8.1 UMTS Charging Architecture (Release 6)

Figure 8.2 illustrates the charging architecture and the information flows for offline and online charging in UMTS R6. Although this chapter will focus on online charging, we still briefly describe the R6 offline charging. Like UMTS R5, in the R6

Charging for Mobile All-IP Telecommunications Yi-Bing Lin and Sok-Ian Sou

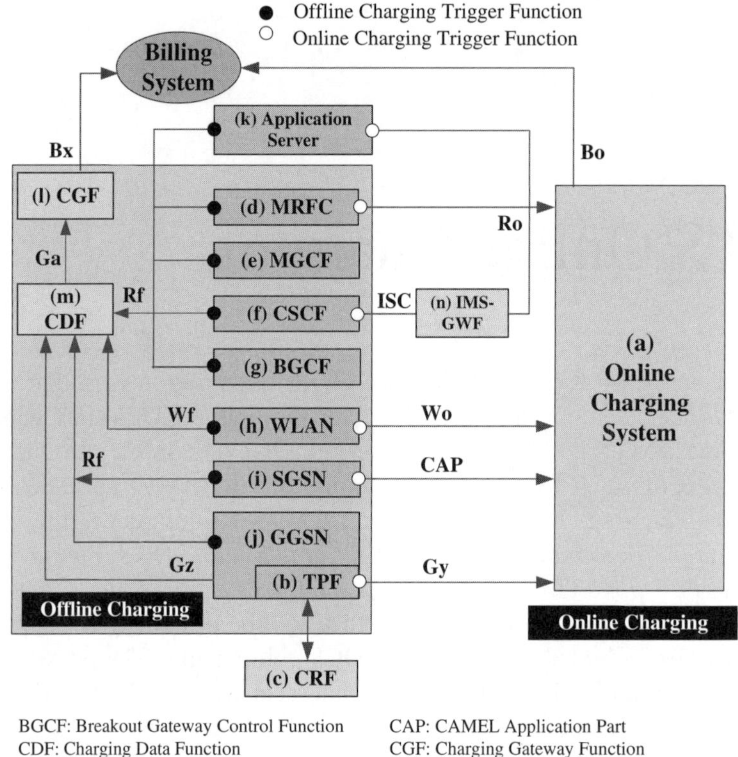

Figure 8.2 UMTS R6 charging architecture

architecture,the offline *Charging Trigger Function* (CTF; see the bullets in Figure 8.2) is a mandatory function integrated in all network nodes (Figure 8.2(d)–(k)) to provide offline charging functionality. The CTF provides metrics that identify the users and the consumption of network resources, and generates chargeable events from these metrics. It then forwards the offline charging information to the *Charging Data Function* (CDF; Figure 8.2(m)) through interfaces Rf, Wf and Gz (implemented by using the Diameter accounting protocol described in Section 5.4). The CDF processes the charging information and constructs it in a well-defined CDR format. The CDR is then transferred to the *Charging Gateway Function* (CGF; Figure 8.2(l)) via the Ga interface (which is implemented by using the GTP' protocol described in Chapter 4). The CGF performs CDR pre-processing including validation, consolidation, reformatting, error handling, persistent CDR storage, CDR routing and filtering, and CDR file management. Then it passes the consolidated offline charging data to the billing system via the Bx interface. The Bx interface supports interaction between a CGF and the billing system for CDR transmissions by a common, standard file transfer protocol (e.g., FTP or SFTP) [3GP06c, 3GP06d]. The CDR files may be transferred in either push or pull mode in the Bx interface. The offline charging for IMS services in R6 is similar to that in UMTS R5 described in Section 7.1, except that the CCF in R5 is replaced by the CDF in R6.

The OCS (Figure 8.2(a)) handles the subscriber account balance, rating, charging transaction control and correlation [3GP06a, 3GP07a]. With the OCS, a telecom operator ensures that credit limits are enforced and resources are authorized on a per-transaction basis. Online charging for SGSN (Figure 8.2(i)) is implemented through the CAP interface described in Section 5.1. Online charging for IMS nodes (Figure 8.2(d)–(g), (k)), WLAN nodes (h), and GGSN (j), is achieved by the interaction with the OCS through interfaces Ro, Wo and Gy (implemented by using the Diameter credit control protocol described in Section 5.5). The IMS-GWF (Figure 8.2(n)) can be regarded as a special kind of SIP application server, which provides protocol translation between IMS Service Control (ISC) and Diameter.

The OCS activities from the IMS nodes are triggered by the online CTF (see the circles in Figure 8.2). This CTF provides similar functionality as the offline CTF described previously, with several enhancements that support online charging; i.e., requesting, granting and managing resource usage permissions according to what the OCS grants or denies. In other words, the online CTF is able to interrupt a service when the user runs out of credit. The online CTF delays the actual service usage until the permission is granted by the OCS. If the permission is not granted, the CTF denies, blocks or cuts the service usage. Such a decision is made based on the rating and the pricing mechanisms in the OCS. The pricing concepts for broadband IP services are essential to the OCS, and the details are described in [Cou03, Fal00, Kou04].

8.1.1 The OCS Architecture

In online charging, network resource usage is granted by the OCS based on the price or the tariff of the requested service and the balance in the subscriber's account. Figure 8.3 shows the OCS architecture defined in 3GPP 32.296 [3GP06a]. The OCS supports two types of *Online Charging Functions* (OCFs), namely the *Session-Based Charging Function* (SBCF) and the *Event-Based Charging Function* (EBCF).

The SBCF (Figure 8.3(a)) is responsible for network bearer and session-based services such as voice calls, GPRS sessions or IMS sessions. The SBCF triggers the session-based charging mode, and controls SIP sessions by appearing as a *Back-to-Back User Agent* (B2BUA) that sends messages to the initiating UA. It also controls non-SIP based bearer systems (GPRS, 2G and other bearer channels), generally through the CAP. The EBCF (Figure 8.3(b)) is responsible for event-based

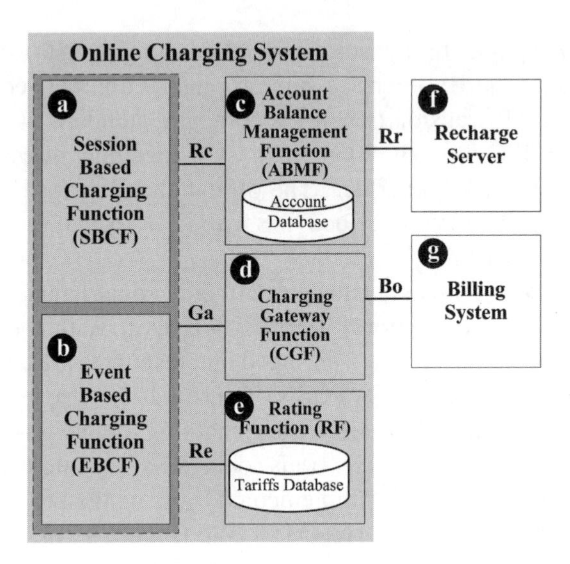

Figure 8.3 The online charging system architecture

services. The EBCF triggers the event-based charging mode, and controls "one-shot" events, such as SMS delivery, and content downloading (e.g., for music or ring tones).

The *Rating Function* (RF; Figure 8.3(e)) determines the price/tariff of the requested network resource usage. The decision about when and how to charge for a session is handled by the charging policies provisioned in the RF. It should be noted that, in some cases, non-chargeable sessions (or sub-sessions) have to be explicitly monitored via "zero rating charging contexts" for consistency. The RF is responsible for providing a cost for the service requested. The SBCF/EBCF furnishes the necessary information (obtained from the IMS/GPRS network nodes) to the RF and receives the rating output (monetary or non-monetary credit units) via the Re interface to be described in Section 8.1.2.

The RF handles a wide variety of ratable instances, including data volume, session/connection time, service events (such as content downloading or message delivery), and combined rating. In addition, different charging rates can be adopted for different time segments (e.g., peak hours and off-peak hours). A real-time RF should be implemented as a high-scale, high-performance engine that enables rapid service deployment without overhauling existing billing systems or replacing otherwise competent legacy systems. The RF implementation should ensure that mediation and charging solutions are not impacted by the latent responses to rating requests that are typical of billing systems not designed to handle the scale of today's services on today's networks. It thus guarantees the best in-session subscriber experience as the network rates sessions and events.

The *Account Balance Management Function* (ABMF; Figure 8.3(c)) maintains user balances and other account data. When a user's credit depletes, the ABMF connects the *Recharge Server* (Figure 8.3(f)) to trigger the recharge account function. The SBCF/EBCF interacts with the ABMF to query and update the user's account. The CDRs generated by the charging functions are transferred to the *Charging Gateway Function* (CGF; Figure 8.3(d)) in real time. The CGF acts as a gateway between the IMS/GPRS network and the billing system (Figure 8.3(g)). Note that both the CGF in Figure 8.3(d) and the CGF in Figure 8.2(l) collect CDRs through the Ga interface. In Figure 8.3, the CGF passes the consolidated online charging data to a billing system via the Bo interface, which is similar to the Bx interface.

8.1.2 Rating Messages for the Re Interface

This subsection elaborates on the Re interface. Specifically, we describe the information exchanged between the charging functions (SBCF and EBCF) and the RF. The RF covers the rating scenarios for all services (GPRS bearer, IMS session and content-based service). The messages and data types used in the Re interface are defined based on the Diameter protocol, and are listed in Table 8.1 [3GP07a].

Table 8.1 The Diameter messages for the Re interface

Message type	Description	Delivery direction
PRQ	Price Request	EBCF → RF
PRS	Price Response	EBCF ← RF
TRQ	Tariff Request	SBCF → RF
TRS	Tariff Response	SBCF ← RF

Details of these messages are described as follows:

- The Price Request/Response message pair determines the price for an event-based service (e.g., short message service). This message exchange can be executed before the service delivery (e.g., for advice of charge and prepaid service) or after service delivery (e.g., for postpaid service).
- The Tariff Request/Response message pair determines the tariff information for a session-based service (e.g., voice calls). Based on the tariff information, the SBCF calculates the amount of credit granted to the network nodes.

The Price Request message contains the following *Attribute-Value Pairs* (AVPs) [3GP06a]:

- The *Session-Id* AVP is used to match the corresponding response message.
- The *Actual-Time* AVP contains the timestamp when the request message is generated.

- The *Subscription-Id* AVP contains the Subscription-Id-Type and Subscription-Id-Data that identify the charged party. The Subscription-Id-Type can be a phone number (type END_USER_E164 with value 0), an IMSI number (type END_USER_IMSI with value 1), a SIP URI (type END_USER_SIP_URI with value 2), a network access identifier (type END_USER_NAI with value 3) or a private identity (type END_USER_PRIVATE with value 4). The Subscription-Id -Data is the charged party's identity (e.g., a mobile phone number +886968311026 for type END_USER_E164).
- The *Service-Identifier* AVP is an Unsigned32 number that identifies the requested service (e.g., IMS call, messaging service, or content downloading).
- The *Destination-Id* AVP contains the destination identifier (e.g., the called party's MSISDN in a call or the recipient's SIP-URI in a message delivery) of the service session/event.
- The *Service-Information* AVP contains the service-specific parameters. For the IMS service, this AVP includes the calling and the called party addresses, the application server information, the IMS charging identifier, and the IMS media information.
- The *Extension* AVP provides additional information to meet operator-specific requirements (e.g., contract parameters).

The Price Response message contains the following AVPs:

- The *Session-Id* AVP is used to match the corresponding request message.
- The *Price* AVP specifies the price of the requested service in the Unsigned32 format.
- The *BillingInfo* AVP contains textual description for bill presentation in the UTF8String format.

The Tariff Request message contains AVPs such as the *Session-Id*, the *Actual-Time*, the *Subscription-Id*, the *Service-Identifier*, the *Destination-Id*, the *Service-Information* and the *Extension* AVPs. These AVPs are the same as those in the Price Request message.

The Tariff Response message contains the following AVPs:

- The *Session-Id*, the *Price* and the *Billing-Info* AVPs are the same as those in the Price Response message.
- The *Monetary-Tariff* AVP contains the current tariff information to indicate the amount of credit charged per time interval or per data usage interval.
- The *Tariff-Switch-Time* AVP indicates the time period (in seconds) from when the Tariff Request message is generated to when the next tariff switch occurs.
- The *Next-Monetary-Tariff* AVP contains the tariff information that is valid after a tariff switch (listed in the *Tariff-Switch-Time* AVP) has occurred.

- The *Expiry-Time* AVP contains the time period (in seconds) from when the Tariff Request message is generated to when the tariff information contained in this message expires.
- The *Valid-Units* AVP contains the maximum amount of credit granted to the service session.

8.2 Online Charging Scenarios

UMTS R6 defines three kinds of online charging: immediate event charging, event charging with unit reservation, and session charging with unit reservation (as introduced in Section 5.5.2). This section describes the detailed message flows for these online charging scenarios.

8.2.1 Immediate Event Charging

In immediate event charging, credit allocation to an IMS node is performed in a single operation, and the credit units are deducted immediately from the subscriber's account. Assume that a subscriber UE1 subscribes to the messaging service with online charging. The message flow for IMS event-based messaging service (offered by an IMS messaging application server) is described below (see Figure 8.4).

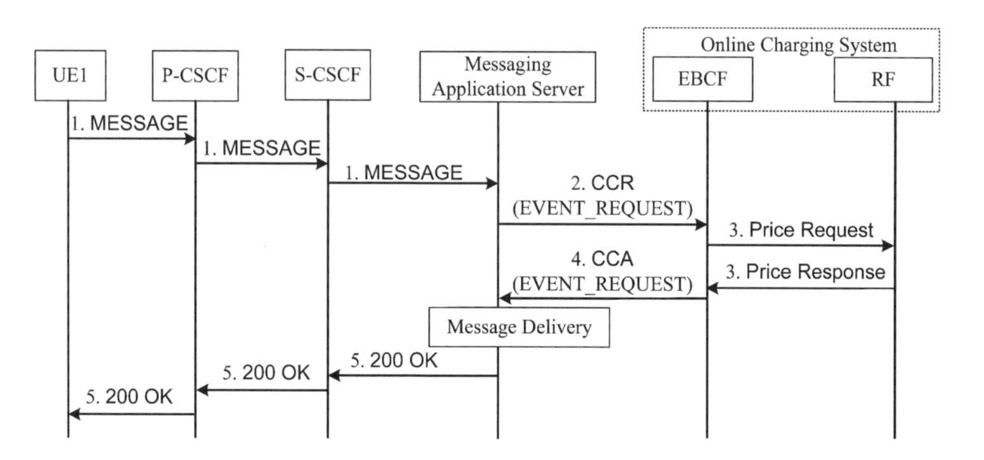

Figure 8.4 The IMS message flow for immediate event charging

Step 1. Subscriber UE1 sends a SIP MESSAGE request to the IMS messaging application server through the P-CSCF and the S-CSCF.

Step 2. The application server sends a CCR message with *CC-Request-Type* "EVENT_REQUEST" and *Requested-Action* "DIRECT_DEBITING" to the EBCF in the OCS.

Step 3. Upon receipt of the credit control request, the EBCF requests account information for the subscriber (such as the billing plan and the subscription profile) from the ABMF. Then the EBCF sends a Price Request message to the RF. The RF determines the price of the service based on the subscriber's billing plan. Specifically, the RF calculates the price for the given service according to the service and subscriber information specified in the request. The calculated price and the billing information are returned to the EBCF through the Price Response message.

Step 4. The EBCF performs credit unit deduction through the ABMF. When the deduction is completed, the EBCF replies to the application server with the CCA message.

Step 5. After the application server has obtained the credit indicated in the CCA message, it delivers the messaging service to UE1. If the service is delivered successfully, the application server sends a SIP 200 OK message to UE1 through the S-CSCF and the P-CSCF. If the credit control fails, an appropriate SIP error message (e.g. 401 Unauthorized or 402 Payment Required) is sent to UE1.

Detailed description of Steps 2 and 4 are also given in Figure 5.7 of Section 5.5.2.

8.2.2 *Event Charging with Unit Reservation*

Event charging with unit reservation conducts reserving credit and returning unused credit for an event-based service. The message flow in Figure 8.5 is described as follows:

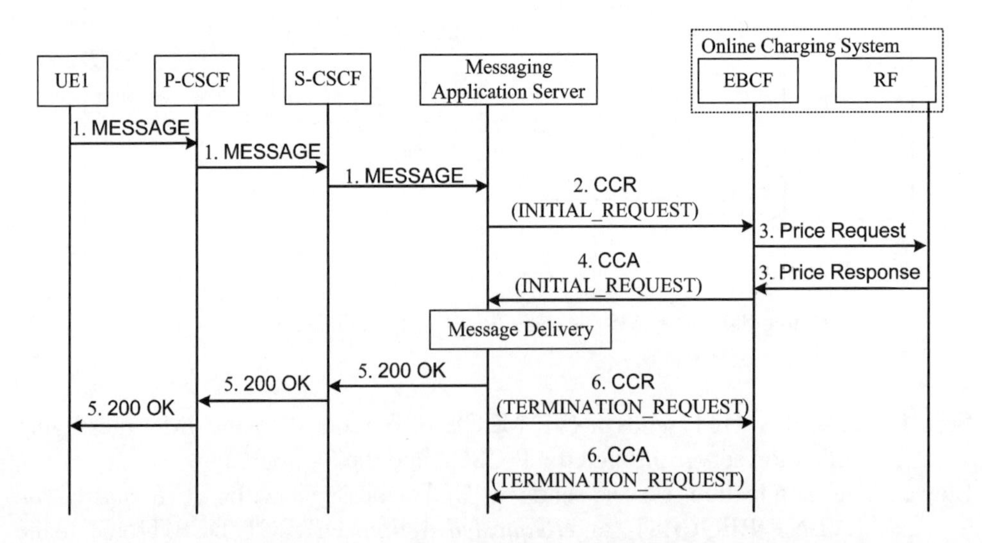

Figure 8.5 IMS message flow for event charging with unit reservation

Step 1. Subscriber UE1 sends a SIP MESSAGE request to the IMS messaging application server through the P-CSCF and the S-CSCF.

Step 2. The application server sends a CCR message with *CC-Request-Type* "INITIAL_REQUEST" to the EBCF of the OCS.

Step 3. Upon receipt of the CCR message, the EBCF requests account information for the subscriber from the ABMF. Then the EBCF sends a Price Request message to the RF. This message includes the subscription identity and the service information. The RF then determines the price of the messaging service. Through the Price Response message, the calculated price and the billing information are returned to the EBCF.

Step 4. The EBCF performs credit unit reservation with the ABMF. Then it replies to the application server with the CCA message indicating the amount of granted credit.

Step 5. After the application server has obtained the credit, the messaging service is delivered to UE1, and then a SIP 200 OK message is sent to UE1 through the S-CSCF and the P-CSCF. If the credit control fails, an appropriate SIP error message (e.g. 401 Unauthorized or 402 Payment Required) is sent to UE1. Assume that the credit control succeeds, Step 6 is executed.

Step 6. After the message is successfully delivered, the application server sends the CCR message with *CC-Request-Type* "TERMINATION_REQUEST" to the EBCF. The EBCF debits the amount of consumed credit from the subscriber's account. Then the OCS replies to the application server with the CCA message.

Detailed description of Steps 2, 4 and 6 are also given in Figure 5.8 of Section 5.5.2.

8.2.3 Session Charging with Unit Reservation

Session charging with unit reservation is performed in credit control of session-based services. Consider the online charging scenario shown in Figure 8.6, where UE1 (with telephone number +886968311026 and SIP URI sip:+886968311026@ims.net1.tw) makes an IMS call to UE2 (with telephone number +886930118839 and SIP URI sip:+886930118839@ims.net2.tw). We assume that UE1 is assigned the IP address 192.168.90.150 and the tariff information for this service is not changed during the session. In Sections 8.2.4 and 8.2.5, we will consider the scenarios where the tariff information is expired during a service session for reasons such as tariff change or QoS change.

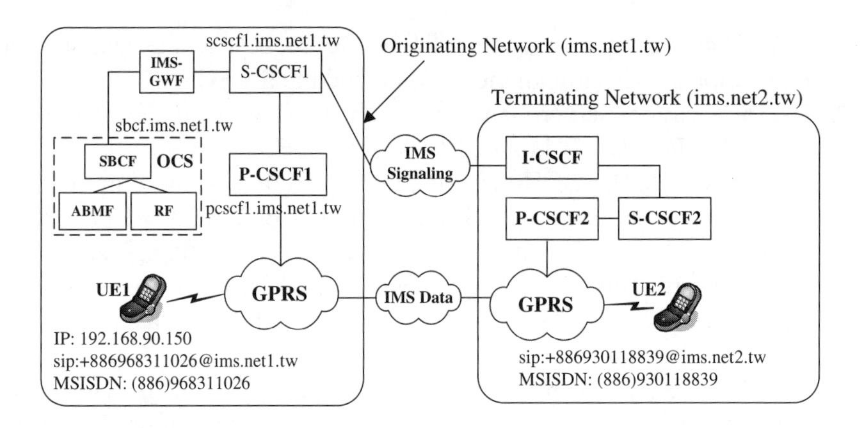

Figure 8.6 The network architecture for an IMS session with online charging

The IMS message flow for session charging with unit reservation (see Figure 8.7) is described as follows (the IMS-GWF between the S-CSCF1 and the OCS is not shown to simplify the description):

Step 1. Subscriber UE1 sends a SIP INVITE request to S-CSCF1 through P-CSCF1. The request message contains the *User-Session-Id* 117524@192.168.90.150, where the number 117524 uniquely identifies the user session for the IP address 192.168.90.150. The message also contains the *IMS-Charging-Identifier* 05110469@pcscf1.ims.net1.tw, where the number 05110469 uniquely identifies the charging event for the IMS call handled by P-CSCF1 (pcscf1.ims.net1.tw). These identities can be generated by any sequence numbers (or random strings) combined with the host name (or IP address) of the node that guarantee to be unique during the session lifetime.

Step 2. S-CSCF1 sends a CCR message with *CC-Request-Type* "INITIAL_REQUEST" to the SBCF of the OCS [IET05]. Details of the CCR message content are listed in Table 8.2. In this example, S-CSCF1 generates a *Session-Id* "4455563c3d@scscf1.ims.net1.tw" to uniquely identify the credit control session (with the number 4455563c3d) between the S-CSCF (scscf1.ims.net1.tw) and the SBCF of the OCS (sbcf.ims.net1.tw). The *CC-Request-Number* "1" is the sequence number of the CCR message (which indicates that the CCR message is the first message in the credit control session). The *Subscription-Id* "886968311026 for type END_USER_E164" is the telephone number of UE1. The *Service-Information* contains the service parameters for the IMS sessions and the *Session Description Protocol* (SDP) parameters for the voice packets [IET02, IET06]. The service context ID is "MNC.MCC.6.32260@3gpp.org" as described in Section 5.5.1.

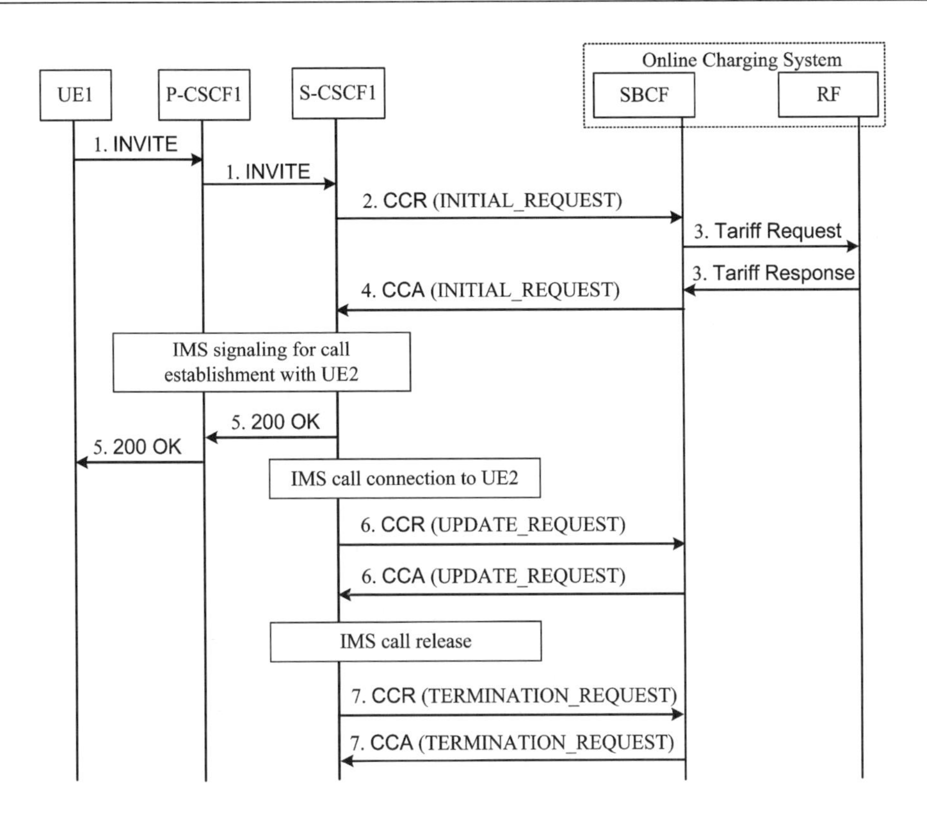

Figure 8.7 The IMS message flow for session charging with unit reservation (the IMS-GWF connecting S-CSCF1 and the OCS is not shown)

Step 3. Upon receipt of the CCR message, the SBCF retrieves the account information and the subscribed QoS profile from the ABMF (Figure 8.3(c)). Then the SBCF sends a Tariff Request message to the RF to determine the tariff of the IMS call. Based on the subscriber information, the RF replies with the Tariff Response message to the SBCF. This message includes the billing plan and the tariff information for the IMS service.

Step 4. When the tariff information is received, the SBCF performs credit unit reservation with the ABMF. Then it replies to S-CSCF1 with the CCA message containing the granted credit (e.g., the number of minutes or bytes allowed). The CCA message content is listed in Table 8.3. In this example, the request is successfully processed with *Result-Code* "DIAMETER_SUCCESS", and the OCS grants 300 time units to the IMS session. In this message, the *Credit-Control-Failure-Handling* AVP determines the action in the Diameter client when the Diameter server is temporarily prevented, e.g., because of network failure. In this example, the action is to terminate the credit control session.

Step 5. After S-CSCF1 has obtained the granted credit units, it continues the call setup to UE2 (through P-CSCF2 and S-CSCF2). Details for the IMS call setup are described in Section 7.1 and are omitted here. When UE2 accepts the call, a SIP 200 OK message is sent to UE1 through S-CSCF1 and P-CSCF1. At this point, the IMS session starts.

Table 8.2 An example of the credit control request message content

AVP	Example
Session-Id	4455563c3d@scscf1.ims.net1.tw
Origin-Host	scscf1.ims.net1.tw
Origin-Realm	ims.net1.tw
Destination-Host	sbcf.ims.net1.tw
Destination-Realm	ims.net1.tw
Auth-Application-Id	Diameter Credit Control Application (value 4)
Service-Context-Id	MNC.MCC.6.32260@3gpp.org
CC-Request-Type	INITIAL_REQUEST
CC-Request-Number	1
Subscription-Id	Type: END_USER_E164 Data: 886968311026
User-Name	886968311026@ims.net1.tw
Service-Information	IMS-Information: Role-of-Node: ORIGINATING_ROLE Node-Functionality: S-CSCF User-Session-Id: 117524@192.168.90.150 Calling-Party-Address: sip:+886968311026@ims.net1.tw Called-Party-Address: sip:+886930118839@ims.net2.tw IMS-Charging-Identifier: 05110469@pcscf1.ims.net1.tw SDP-related parameters (omitted)

Table 8.3 An example of the credit control answer message content

AVP	Example
Session-Id	4455563c3d@scscf1.ims.net1.tw
Result-Code	DIAMETER_SUCCESS (value 2001)
Origin-Host	sbcf.ims.net1.tw
Origin-Realm	ims.net1.tw
Auth-Application-Id	Diameter Credit Control Application (value 4)
CC-Request-Type	INITIAL_REQUEST
CC-Request-Number	1
Credit-Control-Failure-Handling	TERMINATE (value 0)
Multiple-Services-Credit-Control	Granted-Service-Unit: CC-Time: 300

Step 6. During the service session, S-CSCF1 supervises the network resource consumption by deducting the granted credit units. If the granted credit (i.e., the allocated 300 time units) is depleted, the S-CSCF1 sends another CCR message with *CC-Request-Type* "UPDATE_REQUEST" to the SBCF. Through this message, S-CSCF1 also reports the amount of used credit, and requests additional credit for the remaining session. The SBCF deducts the consumed credit and reserves extra credit from the ABMF. Then the SBCF acknowledges S-CSCF1 with the CCA message including the amount of the next reserved credit. Note that this step may repeat several times before the service session is complete.

Step 7. When the service session is complete, S-CSCF1 sends the CCR message with *CC-Request-Type* "TERMINATION_REQUEST" to the SBCF. The SBCF debits the consumed credit units from the subscriber's account with the ABMF. Then the SBCF sends the CCA message to S-CSCF1.

Detailed description of Steps 2, 4, 6 and 7 are also given in Figure 5.9 of Section 5.5.2.

8.2.4 Support of Tariff Change for Session-based Charging

During a service session, the tariff information may be changed when a specified event occurs. For example, assume that the tariff plan for an IMS call is $1 per minute during peak hours (e.g., 8:00am–11:00pm) and $0.50 per minute during off-peak hours (e.g., 11:00pm–8:00am). A user starts an IMS call at 10:55pm and terminates at 11:05pm. The charge for the first five minutes is $5; the charge for the last five minutes is $2.50. In this scenario, the tariff information retrieved from the RF at 10:55pm and that retrieved at 11:00pm are different. Therefore, the S-CSCF needs to reauthorize the granted credit when the valid period has expired at 11:00pm. Figure 8.8 illustrates the message flow for the above scenario, which is described in the following steps:

Step 1. The S-CSCF sends a CCR message with *CC-Request-Type* "INITIAL_REQUEST" to the SBCF of the OCS.

Step 2. Upon receipt of the CCR message, the SBCF requests account information and the subscriber profile from the ABMF. Then the SBCF sends a Tariff Request message to the RF to determine the tariff of the IMS session. The RF replies to the SBCF with the Tariff Response message. This message includes the billing plan and the tariff information based on the service and subscriber information. In this example, the response message also contains the tariff expiry time (with the value TariffSwitchTime1). In the aforementioned IMS call example, the Tariff Request message is received

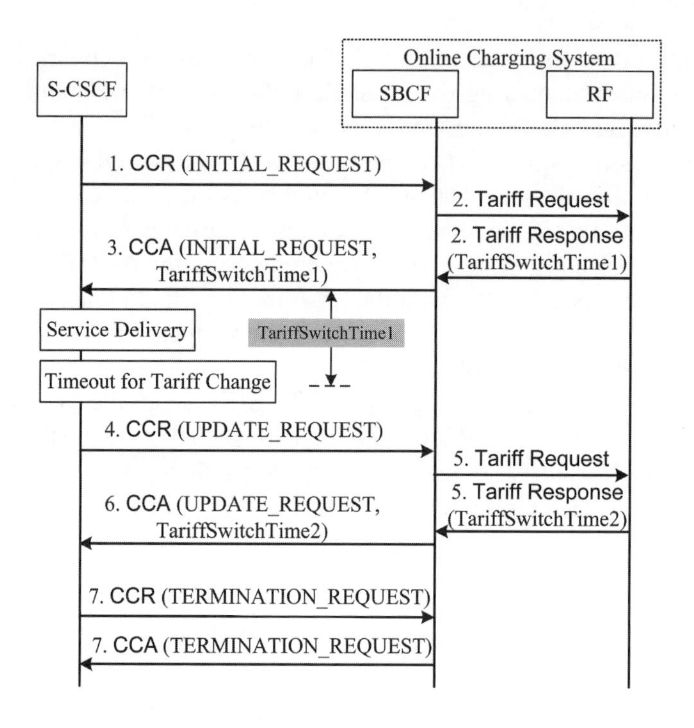

Figure 8.8 IMS message flow for session charging with tariff change

at 10:55pm, and the tariff information expires at 11:00pm. Therefore, TariffSwitchTime1=300 seconds.

Step 3. The SBCF performs credit unit reservation with the ABMF. Then it replies to the S-CSCF with the CCA message containing the granted credit units (e.g. number of minutes) and the expiry time of these granted credit units. After the S-CSCF has obtained the credit units, the IMS session starts.

Step 4. When the timer for the granted credit has expired, the S-CSCF sends another CCR message with *CC-Request-Type* "UPDATE_REQUEST" to the SBCF. The network node reports the amount of used credit, and possibly requests additional credit units. The SBCF deducts the amount of consumed credit units.

Step 5. The SBCF sends a Tariff Request message to the RF to determine the new tariff of the IMS session. The RF replies to the SBCF with the Tariff Response message, which includes new tariff information and the related tariff switch time (e.g., TariffSwitchTime2).

Step 6. Based on the new tariff information, the SBCF reserves extra credit units from the ABMF. The SBCF then acknowledges the S-CSCF by sending the CCA message, which contains the amount of the reserved credit for the service session.

Step 7. When the service session is complete, the S-CSCF sends a CCR message with *CC-Request-Type* "TERMINATION_REQUEST" to the SBCF. The SBCF debits the consumed credit and returns the unused credit units to the subscriber's account in the ABMF. Then the SBCF replies to the S-CSCF with the CCA message.

The reader is encouraged to think about alternatives to connect the OCS with the IMS in this example. For example, can we connect the OCS with the P-CSCF instead of the S-CSCF? Why or why not?

8.2.5 Support of QoS Change for Session-based Charging

During a service session, the tariff information may be expired when the negotiated QoS of the IMS session is changed. For example, due to user movement between different UMTS coverage areas, and depending on the workload of the radio network, the QoS of the IMS session may change from time to time. In this case, the S-CSCF needs to reauthorize the amount of granted credit. Figure 8.9 illustrates the message flow for the above scenario with the following steps:

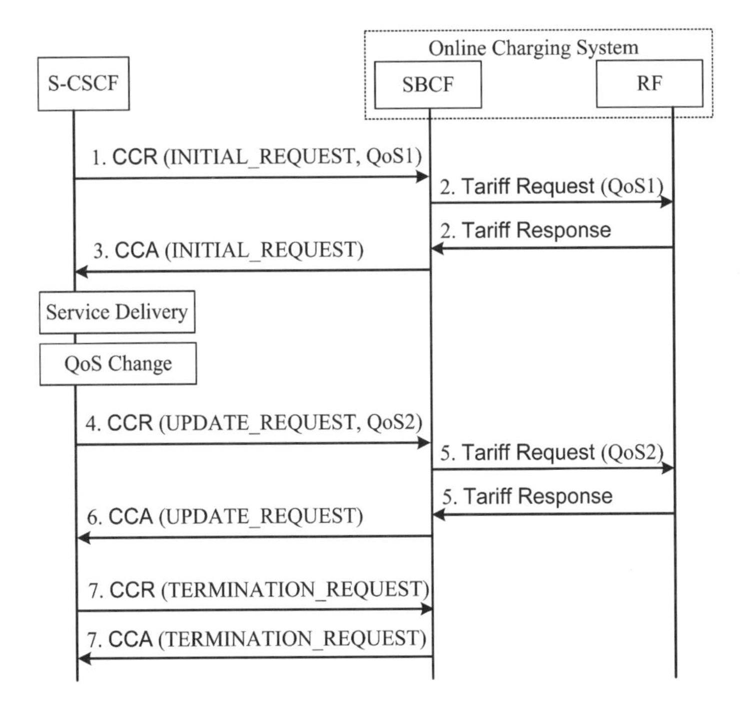

Figure 8.9 IMS message flow for session charging with QoS change

Step 1. The S-CSCF sends a CCR message to the SBCF. This message contains *CC-Request-Type* "INITIAL_REQUEST". Suppose that the related QoS parameter is QoS1 (for 2 Mbps bandwidth).

Step 2. Upon receipt of the CCR message, the SBCF requests account information and the subscriber profile from the ABMF. Then the SBCF sends a Tariff Request message to the RF. The message specifies the requested QoS in the *Service-Information* AVP. The RF replies to the SBCF with the Tariff Response message. This message includes the billing plan and the applicable tariff information for this session.

Step 3. Based on the received tariff information, the SBCF performs credit unit reservation with the ABMF. Then it replies to the S-CSCF with the CCA message, which contains the amount of granted credit (e.g. the number of minutes or bytes allowed) and the trigger event type (i.e., CHANGE_IN_QOS).

Step 4. After the S-CSCF has obtained the reserved credit units, the IMS session starts. When the negotiated QoS of the session changes, the S-CSCF sends another CCR message with *CC-Request-Type* "UPDATE_REQUEST" and the modified QoS parameter (e.g., QoS2 for 384 Kbps bandwidth) to the SBCF. This CCR message includes the *Reporting-Reason* "RATING_CONDITION_CHANGE" and *Trigger-Type* "CHANGE_IN_QOS". The S-CSCF also reports the amount of used credit and possibly requests additional credit units for the remaining session.

Step 5. Upon receipt of the CCR message from the S-CSCF, the SBCF deducts the amount of consumed credit units. Then the SBCF sends the Tariff Request message to the RF to determine the new tariff of the IMS session for new QoS parameter (i.e., QoS2 for 384 Kbps bandwidth). The RF replies to the SBCF with the Tariff Response message including new tariff information.

Step 6. The SBCF reserves extra credit units from the ABMF. The SBCF then acknowledges the S-CSCF with the CCA message including the amount of reserved credit for the service session.

Step 7. When the service session is complete, the S-CSCF sends another CCR message to the SBCF with *CC-Request-Type* "TERMINATION_REQUEST". The SBCF debits the consumed credit units and releases the unused credit from the subscriber's account in the ABMF. Then the SBCF replies to the S-CSCF with the CCA message.

We note that the amount of credit reserved in the CCR/CCA message exchange may affect the performance of online charging. The details are described in Appendices C and D.

8.3 Concluding Remarks

In this chapter, we introduced the UMTS R6 *Online Charging System* (OCS). Specifically, we described the functionalities of the OCS components including the session-based charging function, the event-based charging function, the rating function and the account balance management function. We also elaborated on the contents of the rating messages exchanged between the charging functions and the rating function within the OCS. The reader is encouraged to study the interfaces Rr and Bo described in [3GP05, 3GP06a, 3GP06c, 3GP06d]. Based on the online charging architecture, we also gave several examples to illustrate how the OCS handles credit reservation and deduction in the IMS services. Note that in Step 3 of Sections 8.2.1–8.2.5, and Steps 6 and 7 of Sections 8.2.3–8.2.5, the detailed interaction between the EBCF/SBCF and the ABMF are omitted and will be discussed in Appendix D.

Review Questions

1. Describe the OCS architecture. Which IMS/GPRS nodes will interact with the OCS?
2. Describe the differences between event-based service and session-based service in terms of charging. Can we implement these messages using RADIUS?
3. Describe the rating messages in the Re interface. Can we implement these messages using RADIUS?
4. Give a charging scenario where the calling party is different from the charged party.
5. Describe subscription ID types and destination ID types. What are the differences between them?
6. What is the Expiry-Time AVP included in the Tariff Response message? Why is this AVP not found in the Price Response message?
7. Describe three online charging scenarios. Which scenario fits the video clip application?
8. At Step 3 of immediate event charging, the EBCF and the ABMF interact through the Rc interface. Design the messages exchanged in this interface.
9. At Step 6 of event charging with unit reservation, which components in the OCS are involved in debiting the consumed credit units?
10. During an online service session, in what events do the network nodes (e.g., S-CSCF) need to reauthorize the granted credit with the OCS?

11. Give tariff plan examples for voice call and data services. (Hint: You may check the web pages of your local mobile operators.)
12. Describe the contents of the CCR/CCA messages. Give examples for the QoS parameter specified in a CCR message.
13. Give an example when tariff information is changed during an IMS session. Describe the message flow for session-based charging with tariff change.
14. Describe the message flow for session-based charging with QoS change. Suppose that a mobile operator is allowed to adjust the QoS of service sessions due to system workload. How does the load balancing mechanism affect the charging traffic? Use a performance model to support your claim.
15. Consider a complex multi-session service described in [Alc07], which involves the online purchase of a theater ticket using collaborative IMS services, and which requires a highly flexible charging function provided by the OCS. We assume that booking a ticket from the reservation center is charged based on time zones. Tickets are paid for via a credit card transaction. Sample purchase scenario may unfold as follows:

    - The customer calls the booking center, the customer agent provides a list of theaters with available seats (voice call).
    - The customer asks for details about a theater (continued voice call), the agent pushes back a leaflet (document push).
    - Customer inquires about tariffs (continued voice call), the agent displays a lay-out of the theater room showing pricing areas and vacancies from which the customer interactively selects a seat (document share).
    - Upon validation, credit card payment is proposed to the customer (secured web pages).
    - An electronic ticket is sent to the customer (document push and confirmation mail).
    - Purchase session terminates.

    From the OCS viewpoint, we see the following:

    - The purchase is managed as a main session at the OCS with secondary sub-sessions for the voice call, the document push, the document share, etc.
    - The main session has a time-based charging context. Secondary sessions have "zero rating" charging contexts.

- Charging events are generated from the network at various steps of the purchase. Events are correlated to the main session using identifiers at the OCS.

In this service, the credit card transaction is managed by the booking center via a Mobile Payment Gateway. Please show the interaction between the IMS/Application Server and the OCS for this multi-session service.

References

[3GP05] 3GPP, 3rd Generation Partnership Project; Technical Specification Group Services and Systems Aspects; Telecommunication management; Charging management; Charging data description for the Packet Switched (PS) domain (Release 5), 3G TS 32.215 version 5.9.0 (2005-06), 2005.

[3GP06a] 3GPP, 3rd Generation Partnership Project; Technical Specification Group Service and System Aspects; Telecommunication management; Charging management; Online Charging System (OCS): Applications and interfaces (Release 6), 3G TS 32.296 version 6.3.0 (2006-09), 2006.

[3GP06b] 3GPP, 3rd Generation Partnership Project; Technical Specification Group Core Network; IP Multimedia Subsystem (IMS); Stage 2 (Release 5), 3G TS 23.228 version 5.15.0 (2006-06), 2006.

[3GP06c] 3GPP, 3rd Generation Partnership Project; Technical Specification Group Service and System Aspects; Telecommunication management; Charging management; Charging Data Record (CDR) file format and transfer (Release 6), 3G TS 32.297 version 6.2.0 (2006-09), 2006.

[3GP06d] 3GPP, 3rd Generation Partnership Project; Technical Specification Group Services and System Aspects; Telecommunication management; Charging management; Charging architecture and principles (Release 6), 3G TS 32.240 version 6.4.0 (2006-09), 2006.

[3GP07a] 3GPP, 3rd Generation Partnership Project; Technical Specification Group Service and System Aspects; Telecommunication management; Charging management; Diameter charging applications (Release 6), 3G TS 32.299 version 6.12.0 (2007-09), 2007.

[3GP07b] 3GPP, 3rd Generation Partnership Project; Technical Specification Group Core Network and Terminals; Policy and charging control architecture (Release 7), 3G TS 23.203 version 7.5.0 (2007-12), 2007.

[3GP07c] 3GPP, 3rd Generation Partnership Project; Technical Specification Group Service and System Aspects; Telecommunication management; Charging management; IP Multimedia Subsystem (IMS) charging (Release 6), 3G TS 32.260 version 6.8.0 (2007-03), 2007.

[Alc07] Alcatel-lucent, The Alcatel-lucent 8610 Instant Convergent Charging Suite: Fixed Mobile Convergence Overview, Release 4.5, 2007.

[Cou03] Courcoubetis C. and Weber, R., *Pricing Communication Networks: Economics, Technology, and Modelling*. John Wiley & Sons, Ltd., Chichester, UK, 2003.

[Fal00] Falkner, M., *et al.*, An overview of pricing concepts for broadband IP networks, *IEEE Communications Surveys & Tutorials*, 2nd Quarter: 2–13, 2000.

[IET02] IETF, SIP: Session Initiation Protocol. IETF RFC 3261, 2002.

[IET05] IETF, Diameter Credit-Control Application. IETF RFC 4006, 2005.

[IET06] IETF, SDP: Session Description Protocol. IETF RFC 4566, 2006.

[Kou04] Koutsopoulou, M., *et al.*, Charging, accounting and billing management schemes in mobile telecommunication networks and the Internet, *IEEE Communications Surveys & Tutorials*, **6**(1): 50–58, 2004.

9

Service Data Flow-based Charging

This chapter describes the service data *Flow-Based Charging* (FBC) functionality and architecture defined in UMTS Release 6 (R6). With the FBC, the IP packets belonging to different types of mobile services can be identified and charged by various billing/tariff plans at the GGSN.

We first explain why the FBC was standardized in R6. In UMTS R5, charging for packet-switched service domain (i.e., GPRS) is simply based on the volume of packet data associated with their *Packet Data Protocol* (PDP) contexts [3GP05a]. In GPRS, a mobile user can activate more than one PDP context. The first activated PDP context and the subsequent PDP contexts are referred to as the primary PDP context and the secondary PDP contexts, respectively. The primary and the secondary PDP contexts share the same PDP context information except for the QoS profiles. When an IMS service is delivered through a GPRS bearer session, the operator needs to charge the IMS signaling message and IMS media data separately. The existing solution for R5 activates two PDP contexts. The IMS signaling is transported via the primary PDP context and the IMS media packets are delivered via the secondary PDP context. In this case, the operator charges the transferred data for IMS signaling and IMS media via two PDP contexts with different *tariff plans*. Typically, the IMS signaling is free of charge, and the IMS media data is charged by the connection time or the volume of GPRS data packets. However, as the number of simultaneous service sessions increase, the aforementioned charging solution is not effective because many PDP contexts should be activated to handle multiple service types, which consume extra network resources. This issue is resolved by the FBC in UMTS R6 to be described in this chapter. Another problem of the R5 solution is that it relies on the PDP context mechanism that is only defined in the UMTS/GPRS network. This solution cannot accommodate other radio access technologies such as WLAN and WiMAX [Wimax].

Charging for Mobile All-IP Telecommunications Yi-Bing Lin and Sok-Ian Sou

On the other hand, besides enhancing the functionality of GGSN, UMTS R6 FBC solution also supports other gateway nodes, such as the *Packet Data Gateway* (PDG) for WLAN [3GP06a]. The FBC utilizes the concept of the service data flow specified by the IP packet filters, where the filters are a part of a charging rule. The service data flows distinguish the transferred packets within the same bearer session. Therefore, without the PDP context mechanism, the PDG of the WLAN can also use the specified IP filters to classify different types of service flows. In this chapter, we use the GPRS bearer session (with the PDP context mechanism) as an example to show how FBC works.

9.1 Online Flow-based Charging Architecture

As illustrated in Figure 9.1, the FBC architecture for IMS/GPRS service consists of the *Traffic Plane Function* (TPF; Figure 9.1(a)), the *Application Function* (AF; Figure 9.1(b)) and the *Charging Rules Function* (CRF; Figure 9.1(c)). The AF/CSCF is only involved when the IMS services are delivered through the GPRS session. Other GPRS services, such as WAP, do not require the involvement of the AF. In this architecture, the MS or UE (Figure 9.1(d)) only needs to activate one PDP context with the GPRS network. Data packets belonging to different service flows are transferred via the same bearer session (i.e., the same PDP context). The components in the FBC architecture are described as follows:

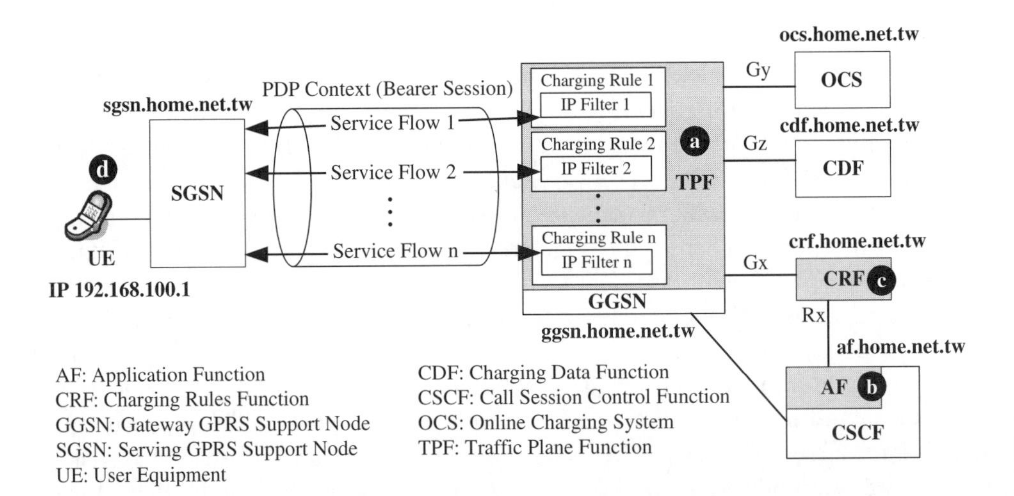

Figure 9.1 FBC architecture for IMS/GPRS services

- Based on the UE identity information, the AF residing in an IMS node (such as P-CSCF or S-CSCF) provides the IMS application information to the CRF through the Rx interface to be described in Section 9.1.3 [3GP07b]. The information includes the application identifier (e.g., for an IMS call), the type of media stream (e.g., audio or video), the IP addresses and port numbers related to the media stream, the requested bandwidth, and the user information.

- The TPF communicates with the CRF to provide bearer session information including the user identity (e.g., the MSISDN), the bearer characteristics (e.g., the negotiated QoS, and the *Access Point Name* (APN)), and the network-related information (e.g., the SGSN address, the SGSN country and network codes). According to the GPRS APN indicated in the PDP context, the TPF requests charging rules from the CRF through the Gx interface [3GP06b]. Based on the installed charging rules, the TPF differentiates the types of user data streams that belong to different service flows within a bearer session. The Gx messages will be described in Section 9.1.1.
- Based on the information received from the TPF and the AF, the CRF configures and selects the appropriate charging rules for the service flows [3GP06b, 3GP07b]. A charging rule contains the charging-related information (such as applying offline or online charging, metering by duration or volume) and an IP filter to classify the transferred packets within the same bearer session. Details of the charging rule will be given in Section 9.1.2.

The flow-based charging can be applied for both offline and online charging. For online charging, the TPF/GGSN requests online credit from the OCS described in Chapter 8. Through the Gy interface implemented by using the *Diameter Credit Control* (DCC) protocol (see Section 5.5), the OCS performs rating and allocates credit depending on the rating group and the traffic characteristics of the TPF. Through the Gz interface implemented by the Diameter protocol (see Section 5.4), the TPF sends the offline CDRs to the *Charging Data Function* (CDF) as described in Chapter 8. In the following subsections, we describe the Gx and the Rx interfaces. These two interfaces are also referred to as the Diameter Rx application and the Diameter Gx application, respectively, which are different from the user data application (such as WAP or streaming service) discussed in this book. The related message flows for charging rule handling will be elaborated in Section 9.2.

9.1.1 Messages for the Gx Interface

The Gx interface (between the CRF and the TPF) provisions the charging rules to the service flows. This interface utilizes the DCC protocol [IET05b], where the CRF acts as a DCC server and the TPF acts as a DCC client. Table 9.1 lists the message types and the delivery direction for the Gx Interface. Note that the message types in the Gx interface contain *Attribute-Value Pairs* (AVPs) different from those in the Rf interface mentioned in Chapter 5. The details for the Gx messages are described below.

Table 9.1 The Gx Diameter messages

Message type	Description	Delivery direction
CCR	Credit Control Request	TPF → CRF
CCA	Credit Control Answer	TPF ← CRF
RAR	Re-Auth Request	TPF ← CRF
RAA	Re-Auth Answer	TPF → CRF

When a bearer session is established (i.e., the GPRS PDP context is activated), the TPF sends an INITIAL_REQUEST Credit Control Request (CCR) message to the CRF. This message contains the service information such as the type of radio access technology, the UE's IP address, the QoS profile negotiated, and the APN. The CRF determines the charging rules and then sends them to the TPF by using the Credit Control Answer (CCA) message. Both the CRF and the TPF maintain information for the credit control session. When an existing bearer session is modified (e.g., the GPRS PDP context is updated with new QoS), the TPF sends an UPDATE_REQUEST CCR message to the CRF. This message contains the new service information. The CRF determines the new charging rules and then sends them to the TPF by using the CCA message. Note that these credit control messages carry the same session ID to uniquely identify the credit control session (for the purpose of requesting charging rules). In order to distinguish the credit control sessions for requesting online credit (described in Chapter 8) from that for requesting charging rule, we use the "charging rule request session" to specifically represent the credit control session between the CRF and the TPF. The CCR message for the charging rule request session contains the following AVPs:

- The *Session-Id* AVP identifies the charging rule request session.
- The *Origin-Host* and the *Origin-Realm* AVPs contain the IP address (or domain name) and the realm of the TPF. In Figure 9.1, the Origin-Host is ggsn.home.net.tw and the Origin-Realm is home.net.tw.
- The *Destination-Host* and the *Destination-Realm* AVPs contain the IP address (or domain name) and the realm of the CRF. In Figure 9.1, the Destination-Host is crf.home.net.tw and the Destination-Realm is home.net.tw.
- The *Auth-Application-Id* AVP contains the application identifier for the Diameter Gx application, which is 16777224 allocated by the *Internet Assigned Numbers Authority* (IANA).
- The *CC-Request-Type* AVP indicates the request type for the charging rule, which is the same as that described in Section 5.5.1.
- The *CC-Request-Number*, the *Subscription-Id* , and the *Termination-Cause* AVPs are the same as those described in Section 5.5.1.
- The *3GPP-RAT-Type* AVP indicates the radio access technology (e.g., UTRAN, WLAN, WiMAX, and so on) that is currently serving the UE.

- The *User-Equipment-Info* AVP contains the identification and capabilities of the terminal; e.g., *International Mobile station Equipment Identity* (IMEI) and *Software Version* number (IMEISV).
- The *3GPP-GPRS-Negotiated-QoS-Profile* AVP contains the negotiated QoS profile retrieved from the activated PDP context.
- The *3GPP-SGSN-MCC-MNC* AVP contains the *Mobile Country Code* (MCC) and the *Mobile Network Code* (MNC) of the SGSN.
- The *3GPP-SGSN-Address* AVP contains the address of the SGSN involved in this bearer session. In Figure 9.1, the 3GPP-SGSN-Address is sgsn.home.net.tw.
- The *Framed-IP-Address* AVP contains the IP address allocated to the user. In Figure 9.1, the UE is assigned an IP address 192.168.100.1.
- The *Called-Station-ID* AVP contains the address of the destination where the user is connected to (e.g., the APN for a GPRS service, the called party address for an IMS call, and so on.)
- The *Bearer-Usage* AVP indicates how the bearer is used; for example, "GENERAL" (with value 0) indicates that no specific bearer usage information is available; "IMS_SIGNALLING" (with value 1) indicates that the bearer is used for IMS signaling.
- The *TFT-Packet-Filter-Information* AVP contains the GPRS *Traffic Flow Template* (TFT) information associated with a PDP context. This AVP includes the type of service, the precedence and the IP packet filters related to the bearer session (i.e., the source and the destination IP addresses and port numbers, the direction for the IP flows). In Figure 9.1, the source IP address is 192.168.100.1 with any port number.

The CCA message acknowledges a CCR message. The CCA message contains the following AVPs:

- The *Session-Id*, the *Origin-Host*, the *Origin-Realm*, the *Auth-Application-Id*, the *CC-Request-Type*, the *CC-Request-Number* AVPs are similar to those in the CCR message.
- The *Result-Code* AVP contains the result of the charging rule request. For example, DIAMETER_SUCCESS indicates a successful message handling (see Table 5.7).
- The *Event-Trigger* AVP specifies the events that will trigger the TPF to re-authorize the charging rules. The events can be change of SGSN (with value 0), change of QoS (with value 1), change of radio access technology (with value 2), change of traffic flow template (with value 3), and change of *Public Land Mobile Network* (PLMN; with value 4). When an event occurs, the TPF should send a CCR message to request new charging rules.
- The *Charging-Rule-Install* AVP is used to activate, install or modify the charging rules. This message includes the unique charging rule names and the related information regarding the service flows. More details will be described in Section 9.1.2.

- The *Charging-Rule-Remove* AVP is used to deactivate or remove the charging rules.
- The *Charging-Information* AVP contains the OCS address for online charging and the CDF address for offline charging.

The Re-Auth Request (RAR) message informs the TPF to re-authorize the existing charging rules. This message includes the *Session-Id*, the *Origin-Host*, the *Origin-Realm*, and the *Auth-Application-Id* AVPs. These AVPs are similar to those in the CCR message. In Figure 9.1, the OCS address is ocs.home.net.tw, the CDF address is cdf.home.net.tw.

The Re-Auth Answer (RAA) message acknowledges the RAR message. After the RAA message is sent, the TPF should send a CCR message to the CRF to re-authorize the existing charging rules. The RAA message includes the following AVPs:

- The *Session-Id*, the *Origin-Host*, the *Origin-Realm*, the *Auth-Application-Id*, the *Result-Code* AVPs are similar to those in the CCA message.
- The *Experimental-Result* AVP contains the execution result for the RAR message. As defined in the Gx application [3GP06b], the value contains in this AVP may be "Success" or "Permanent Failure".

9.1.2 FBC Charging Rules

In the FBC architecture, the charging rules sent from the CRF to the TPF contain information for handling content-based charging. A charging rule consists of the following AVPs:

- The *Charging-Rule-Name* AVP uniquely identifies the charging rule applied to the service flow handled in the TPF.
- The *Flow-Description* AVP determines the IP filter (i.e., the protocol, the source and the destination IP addresses and port numbers) associated with the service flow. The protocol can be TCP or UDP.
- The *Service-Identifier* AVP identifies the service (such as WAP, video streaming service, and so on) related to the charging rule.
- The *Rating-Group* AVP contains the information (i.e., the charging key) for the TPF to request credit from the OCS. Specifically, the OCS uses the charging key to determine the rating (tariff) for the service flow.
- The *Metering-Method* AVP specifies the parameter to be metered for a service flow. The parameter can be the connected time of the service flow (DURATION), the packet volume transmitted in the service flow (VOLUME), or both duration served and packet volume transmitted within the service flow (DURATION_VOLUME). Table 9.2 lists the values for the Metering-Method AVP.

- The *Precedence* AVP defines the precedence of a charging rule (if multiple charging rules exist). A charging rule with a small precedence value takes the priority over a charging rule with a large precedence value.
- The *AF-Charging-Identifier* AVP contains the charging identifier of the AF. This information is used for charging correlation with the bearer layer. Specifically, this AVP for AF/CSCF contains the *IMS Charging Identifier* (ICID) described in Section 5.6.2.
- The *Flows* AVP contains the IP flow identifiers. This identifier contains the media component number (i.e., the ordinal number of the media component listed in the SDP) and the flow number (i.e., the ordinal number of the IP flow assigned for the media component [3GP05b]). This AVP is used in charging correlation with the AF Charging Identifier.

Table 9.2 The Metering-Method AVP

Description	Value
DURATION	0
VOLUME	1
DURATION_VOLUME	2

9.1.3 Messages for the Rx Interface

The Rx interface (between the AF and the CRF) provisions IMS application information for the service flows. The Rx interface is implemented based on the Diameter *Network Access Server Requirements* (NASREQ) Protocol [IET05a], where the CRF acts as a Diameter server and the AF acts as a Diameter client. Table 9.3 lists the Rx message types and the delivery direction. Details of these messages are described below.

Table 9.3 The Rx Diameter messages

Message type	Description	Delivery direction
AAR	AA Request	AF → CRF
AAA	AA Answer	AF ← CRF
RAR	Re-Auth Request	AF ← CRF
RAA	Re-Auth Answer	AF → CRF

The AA Request (AAR) message provides IMS application information. This message contains the following AVPs:

- The *Session-Id*, the *Origin-Host*, the *Origin-Realm*, the *Destination-Realm* and the *Subscription-ID* AVPs are similar to those in the CCR message. In Figure 9.1,

the Origin-Host is af.home.net.tw and the Origin-Realm is home.net.tw. The Destination-Host is crf.home.net.tw and the Destination-Realm is home.net.tw.
- The *Auth-Application-Id* AVP contains the Diameter Rx application identifier, which is 16777229 allocated by the IANA.
- The *Media-Component-Description* AVP contains service information for a media component within an IMS session. Specifically, it includes the media type (audio, video, data, application, control, text, message, etc.) and the requested bandwidth in bits per second. This AVP is used in the GGSN to determine the negotiated QoS for the GPRS session and to select the charging rules.
- The *AF-Charging-Identifier* AVP contains the charging identifier of the AF. This information is used for charging correlation with the bearer layer, which is also recorded in the charging rule.
- The *Specific-Action* AVP specifies the events that the CRF needs to report to the AF. The events can be bearer establishment (with value 5), bearer release (with value 4), and so on. If one of the specified events occurs, the CRF should send the Re-Auth Request message to the AF (to be described later).

An AA Answer (AAA) message is the acknowledgment of an AAR message. The AAR message includes the *Session-Id*, the *Origin-Host*, the *Origin-Realm*, the *Auth-Application-Id*, the *Result-Code* and the *Experimental-Result* AVPs. These AVPs are similar to those in the RAR message described in Section 9.1.1.

An RAR message is used to inform the AF that a specific event (e.g., the bearer establishment or the bearer release) has occurred. Note that the RAR/RAA messages used in the Rx and the Gx interfaces contain different AVPs.

The RAR message used in the Rx interface contains the following AVPs:

- The *Session-Id*, the *Origin-Host*, the *Origin-Realm*, the *Destination-Realm*, the *Auth-Application-Id*, the *Subscription-ID*, and the *Specific-Action* AVPs are similar to those in the AAR message.
- The *Flows* AVP contains the flow identifiers that are affected in this session.

The RAA message acknowledges an RAR message. The RAA message includes the *Session-Id*, the *Origin-Host*, the *Origin-Realm*, the *Auth-Application-Id*, the *Result-Code* and the *Experimental-Result* AVPs. These AVPs are similar to those in the RAR message described in Section 9.1.1.

9.2 Content-based Service for Online TPF/GPRS

This section illustrates how the TPF differentiates the service flows for various content-based services transferred via the same PDP context. We give an online GPRS content-based charging example.

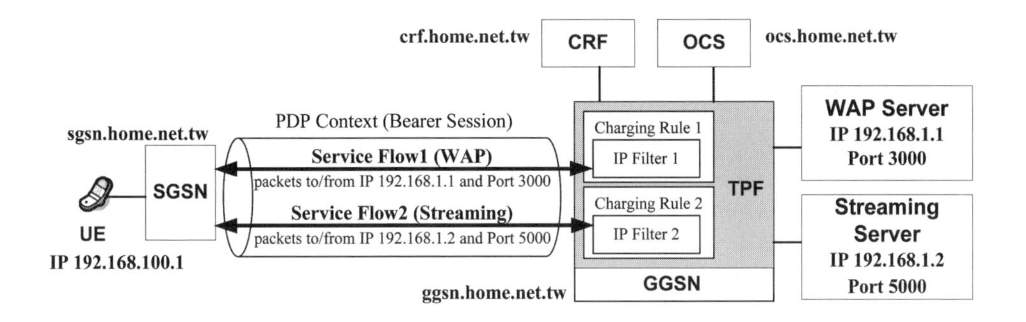

Figure 9.2 An example of service data flows

The WAP service and the video streaming service illustrated in Figure 9.2 are delivered within a bearer session (i.e., the services are specified by the same PDP context). Two charging rules are configured for these two content-based services: Charging rule 1 is applied to the WAP service flow. Charging rule 2 is applied to the streaming service flow. The IP packet filters are contained in the charging rules with the source and the destination IP addresses and port numbers. The TPF/GGSN uses the IP filters to screen and log the transferred packets and the duration times for the service flows.

When a user subscribes to the mobile services, she can choose her tariff plans according to her personal requirements and monthly usage. Different kinds of tariff plans are applied to different groups of users (such as family, community, enterprise, and so on). In this example, we assume that the user subscribes to the tariff plan of a group named "Group A". The related tariff plan of Group A is listed in Table 9.4, where the WAP service is charged by the amount of data transmission ($0.10 per KB), and the video streaming service is charged by the viewing time ($0.50 per minute). Details of the *pricing* concepts for broadband IP services are described in [Cou03, Fal00, Kou04].

Table 9.4 A tariff plan example for content-based services

Service type	Method	Tariff of Group A
WAP	Volume-based	$0.10/KB
Video streaming	Time-based	$0.50/minute

In Figure 9.2, assume that a user (i.e., a UE) starts both the WAP service and the streaming service at 11:00pm and terminates them at 11:30pm. In this period, the WAP server has transmitted 80 KB data to the UE, and the streaming server has delivered the video content to the UE for 30 minutes. Details of service information are listed in Table 9.5. According to the tariff plans, the charges for the WAP service and the streaming service are $8 and $15, respectively. Therefore, the user is charged for $23 in total for the services delivered during the period 11:00pm–11:30pm.

Table 9.5 Service information for a content-based service example

Service Type: WAP Service	Service Type: Video Streaming Service
Node Address: 192.168.1.1	Node Address: 192.168.1.2
Port: 3000	Port: 5000
Start Time: 11:00pm	Start Time: 11:00pm
End Time: 11:30pm	End Time: 11:30pm
Data Size: 80 KB	Data Size: 100 KB
Duration: 30 minutes	Duration: 30 minutes

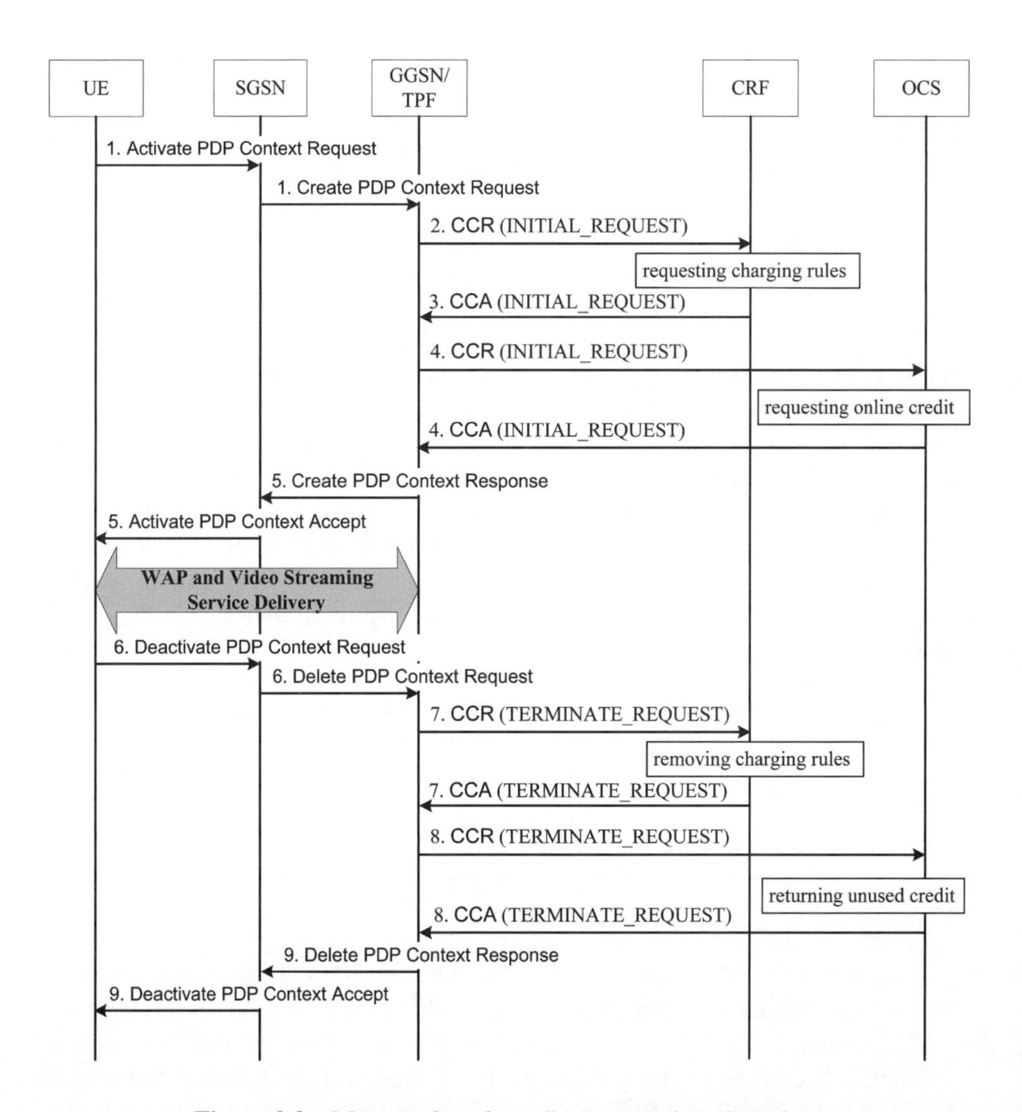

Figure 9.3 Message flow for online content-based services

The message flow for the above example is illustrated in Figure 9.3, and is described as follows:

Step 1. At 11:00pm, the UE initiates the PDP context activation procedure (through the SGSN and the GGSN) to establish a PDP context with a requested QoS profile.

Step 2. The GGSN creates the PDP context record and assigns the IP address 192.168.100.1 to the UE. Then the TPF/GGSN sends a CCR message to the CRF (at address crf.home.net.tw) indicating that a GPRS bearer session is established. The CCR message contains the following information:

- The network-related information includes the *3GPP-RAT-Type* with value "UTRAN" and the *3GPP-SGSN-Address* indicating the SGSN address sgsn.home.net.tw.
- The bearer characteristics include the *3GPP-GPRS-Negotiated-QoS-Profile* with the QoS parameters for the PDP context, the *Called-Station-ID* with the IP address of the WAP server (192.168.1.1:3000 in Figure 9.2) and the streaming server (192.168.1.2:5000 in Figure 9.2), and the *Bearer-Usage* with value "GENERAL".
- The user identity information includes the *Subscription-ID* with the user's IMSI, the *Framed-IP-Address* with value "192.168.100.1" and the *User-Equipment-Info* with the IMEISV of the UE.

Step 3. Based on the received information, the CRF determines the charging rules to be provisioned for the TPF. The CRF sends the CCA message to the TPF. In this example, the message includes two charging rules in the *Charging-Rule-Install* AVP and the OCS address (ocs.home.net.tw) in the *Charging-Information* AVP. The charging rules applied to this example are listed in Table 9.6, and are described as follows:

- The *Charging-Rule-Name* AVPs specify the names for the charging rules, which are "rule 1" and "rule 2", respectively.
- The *Service-Identifier* AVPs specify that these two charging rules are applied to the WAP service and the streaming service, respectively.
- The *Flow-Description* AVPs specify the IP filters for the service flows. The configuration of the IP filters permits the packets of the UE (IP address 192.168.100.1 with any port number) and the packets of the WAP server (192.168.1.1:3000) and the streaming server (192.168.1.2:5000) to pass through the TPF/GGSN.
- The *Online* and *Offline* AVPs indicate that the TPF should enable online charging with the service flows.

Table 9.6 Charging rule examples for content-based services

Charging-Rule-Name: rule 1
Service-Identifier: WAP Service
Flow-Description: (IP filter 1)
permit in ip from 192.168.100.1 to 192.168.1.1:3000
permit out ip from 192.168.1.1:3000 to 192.168.100.1
Online: ENABLE_ONLINE (with value 1)
Offline: DISABLE_OFFLINE (with value 0)
Metering-Method: VOLUME (with value 1)
Rating-Group: Group A

Charging-Rule-Name: rule 2
Service-Identifier: Video Streaming Service
Flow-Description: (IP filter 2)
permit in ip from 192.168.100.1 to 192.168.1.2:5000
permit out ip from 192.168.1.2:5000 to 192.168.100.1
Online: ENABLE_ONLINE (with value 1)
Offline: DISABLE_OFFLINE (with value 0)
Metering-Method: DURATION (with value 0)
Rating-Group: Group A

- The *Metering-Method* AVPs specify that the WAP service is charged by the transferred packet volume and the video streaming service is charged by the duration time.
- The *Rating-Group* AVPs indicate that the tariff plan of Group A is applied to the UE.

Step 4. Upon receipt of the charging rules from the CRF, the TPF configures the IP filters such that the service flows for the WAP and the video streaming services can pass through the TPF/GGSN. To handle online charging, the TPF sends the INITIAL_REQUEST CCR message to the OCS to request online credit. This CCR message indicates that the UE belongs to Group A. According to the tariff plan of Group A (listed in Table 9.4), the OCS calculates the reserved credit for these two content-based services. After the reservation, the OCS replies the TPF with the CCA message to indicate the amount of the granted credit.

Step 5. At this point, the PDP context is successfully activated for the UE. The WAP and the streaming services are delivered to the UE. The TPF counts the transferred packet volume for the WAP service, and calculates the duration time for the streaming service.

Step 6. At 11:30pm, the UE terminates the GPRS session by deactivating the PDP context from the GPRS network.

Step 7. By sending the TERMINATION_REQUEST CCR message to the CRF, the TPF closes the charging rule request session. After the CRF has processed the message, it sends the CCA message to acknowledge the TPF.

Step 8. The TPF closes the online charging session with the OCS by sending the TERMINATION_REQUEST CCR message, which specifies the consumed credit of the services. In this example (see Table 9.5), the consumed credit is 80 credit units (80 KB) for the WAP service and 30 credit units (30 minutes) for the streaming service. The OCS calculates the total charge (i.e., $8 + $15 = $23) for these services. Then the OCS debits the user account and acknowledges the TPF by sending the CCA message.

Step 9. The GGSN releases the PDP context with the UE. The SGSN informs the UE that the PDP context is successfully deactivated.

9.3 Online IMS Flow-based Charging

This section describes how flow-based charging is applied to an IMS service scenario. We note that in Chapters 7 and 8, IMS charging is exercised between the CSCF and the OCS. In such solution, the GGSN does not have the application-level charging knowledge. Therefore, the GGSN does not need to interact with the OCS, and is only responsible for delivering user data at the bearer level.

Through the FBC structure described in this chapter, the online charging for IMS can be handled by the GGSN. In this solution, the GGSN is modified to include the TPF so that it can access the charging rules from the CRF. In other words, the GGSN has the knowledge of IMS media packets and can perform IMS credit control.

Depending on the mobile network configurations, an operator can decide whether to exercise IMS charging on the CSCF or on the TPF/GGSN. In these two online charging solutions, the IMS CDRs are generated by the charging functions of the OCS (the SBCF or the EBCF described in Section 8.1.1).

Figure 9.4 shows an IMS video call scenario through the FBC solution. In the IMS signaling, the SIP/SDP message carries different types of media information. When

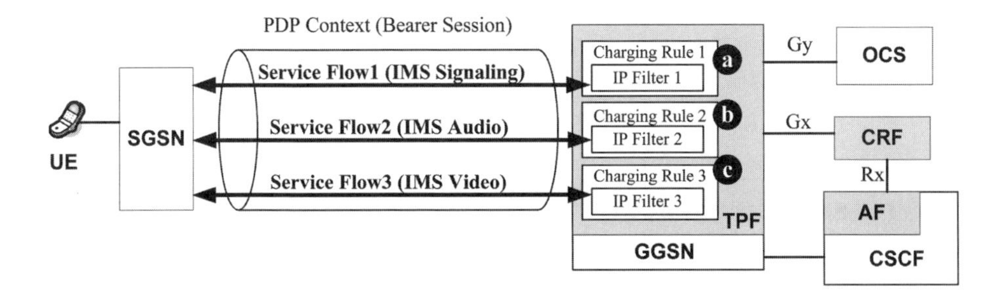

Figure 9.4 An IMS service example with three service data flows

the UE sends a SIP message to the CSCF, the AF implemented in the CSCF provides the media service flow information to the CRF through the Rx interface. Therefore, the IMS media packets can be accurately classified and charged. Through the Gx interface, the CRF then sends the service flow information to the TPF/GGSN, including the IP flow details and the related charging key (i.e., the rating group name in the OCS). For online charging, the TPF/GGSN sends the credit request to the OCS.

In Figure 9.4, there are three service flows transferred via a PDP context: the IMS signaling packets (Service Flow1), the IMS audio packets (Service Flow2), and the IMS video packets (Service Flow3). Figure 9.5 illustrates the message flow for this scenario with the following steps:

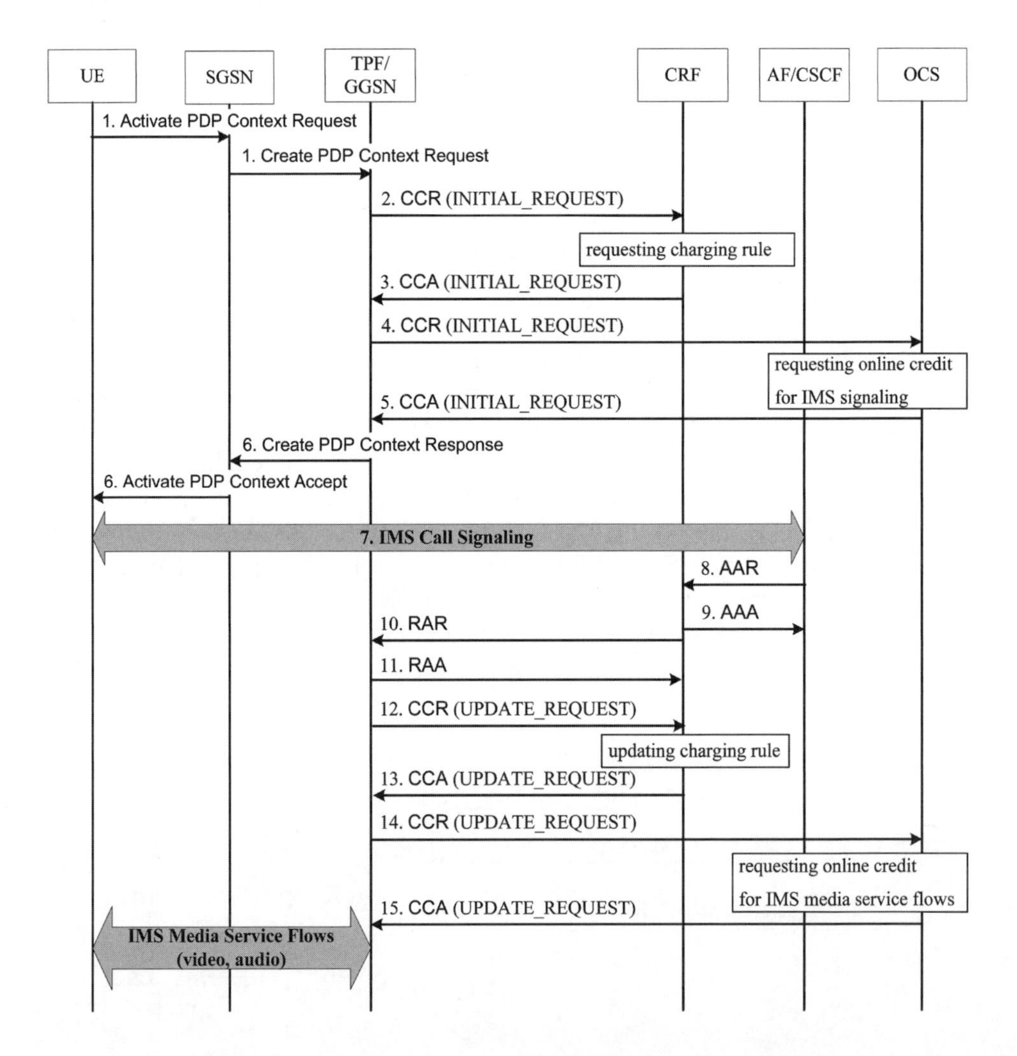

Figure 9.5 Message flow for TPF-based online IMS service

Step 1. The UE initiates the PDP context activation procedure with the GPRS network to establish a PDP context.

Step 2. The GGSN creates the PDP context record. Then the TPF sends a CCR message to the CRF to indicate that a GPRS bearer session of the UE is established.

Step 3. Based on the GPRS session information, the CRF determines the signaling charging rule to be provisioned for the TPF. The CRF then replies the CCA message to the TPF. This message contains the charging rule for IMS signaling packets delivered between the UE and the P-CSCF (Figure 9.4(a)). Assuming that the charging rule indicates that the IMS signaling service flow is handled by online charging, where the related rating group is "Group B" illustrated in Table 9.7.

Step 4. To request the online credit for the IMS signaling transmission, the TPF sends the INITIAL_REQUEST CCR message to the OCS. This CCR message indicates that the *Rating-Group* is "Group B" and the *Service-Identifier* is "IMS Signaling".

Step 5. According to the tariff plan of Group B listed in Table 9.7, the OCS determines that the IMS signaling packets are free of charge. Through the INITIAL_REQUEST CCA message, the OCS then informs the TPF that the service flow for IMS signaling has unlimited quota.

Step 6. The TPF configures the IP filter such that the service flow for IMS signaling can pass through the TPF/GGSN. Then the PDP context is successfully activated for the UE.

Step 7. The UE establishes the IMS call by sending the SIP INVITE message to the P-CSCF. The P-CSCF forwards the message to the related S-CSCF. Details for the IMS call setup are described in Section 7.1 and are omitted here.

Table 9.7 A tariff plan example for IMS video call

Service type	Method	Tariff of Group B
IMS signaling	—	Free
IMS audio	Time-based	$1/minute
IMS video	Time-based	$0.5/minute

Note that we purposely charge the audio and the video separately in this example. In a video call, the video may not be available during some periods of the call duration (e.g., due to the UE mobility). In these periods, the user will only be charged for $1/minute (instead of $1.5/minute).

Step 8. After the call has been established, the AF/CSCF sends the AAR message to the TPF/GGSN. This message provides the IMS session information by including media information for the IMS call, such as the application identifier (i.e., IMS video call), the types of stream (i.e., audio and video), the IP address and the port of UE that will be used to transfer the media packets.

Step 9. The CRF identifies and installs the charging rules for IMS audio and video. Then it acknowledges the AF/CSCF with the AAA message.

Step 10. According to the application information provided from the AF/CSCF, the CRF sends the RAR message to the TPF to request charging rule re-authorization. The purpose of this re-authorization is to add the charging rules for the video and the audio service flows.

Step 11. The TPF acknowledges the CRF through the RAA message.

Step 12. The TPF sends the UPDATE_REQUEST CCR message to the CRF to re-authorize the charging rules. Based on the new information, the CRF determines the charging rules for the IMS audio service flow and the video service flow.

Step 13. Through the CCA message, the CRF downloads the appropriate charging rules to the TPF (Figure 9.4(b) and (c)).

Step 14. When the TPF receives the audio and the video charging rules, it requests online credit from the OCS through the UPDATE_REQUEST CCR message. Note that the CCR message indicates that the *Rating-Group* is "Group B" listed in Table 9.7. The *Service-Identifiers* are "IMS Video" and "IMS Audio". The OCS calculates the reserved credit for the IMS audio and video service flows according to the tariff plan of Group B. In this example, both the IMS audio and video service flows are charged by time duration. The tariff rates for these service flows are $1 and $0.50 per minute, respectively.

Step 15. When the reservation is finished, the OCS replies to the TPF with the CCA message to indicate the amount of granted credit. At this point, the IMS call (including the audio and the video streams) can be delivered to the UE. The TPF allows the IMS media packets to pass through the GGSN. It also calculates the time period for these service flows.

When the IMS call is finished, the TPF sends the charging information to the OCS. The OCS deducts the consumed credit units from the subscriber's account based on the tariff of different service flows. In the above example, we assume that the IMS signaling and the IMS media packets are transferred within the same PDP context and QoS profile. The reader is encouraged to redraw the message flow such that the IMS signaling and the IMS media are transferred through two individual PDP contexts (i.e., with different QoS profiles). In this case, the UE needs to activate the secondary PDP context.

9.4 Policy and Charging Control Integration

Today, the network resources consumption varies considerably among subscribers generally charged the same flat rate for Internet access, where telecom operators are not able to guarantee QoS for all subscribers. Therefore, it is important that a telecom operator can provide different QoSs for subscribers who are charged by different rates. The QoS control in IMS/GPRS is realized by the *Session-Based Local Policy* (SBLP) functionality defined in 3GPP R5 [3GP05c]. The SBLP utilizes the *Policy Decision Function* (PDF; see Figure 9.6(d)) to make policy decisions based on session and media-related information obtained from the CSCF. In other words, the QoS policy is controlled by the PDF, where the policy rules can be formulated based on static information (such as the subscription profile), dynamic information, and the available resources. The combination of such policy rules, once met for a service request, can trigger a desired action (such as allowing the service with the requested bandwidth). This policy rule framework allows the telecom operators to deploy service logic while optimally utilizing the network resources. Specifically, by configuring the policy stored in the PDF, telecom operators are able to implement the QoS policy control flexibly for different applications in various IP networks.

When the QoS policy control and the FBC functionalities described in Section 9.1 are used as separate mechanisms, it will increase the interworking cost between network nodes (e.g., GGSN and CSCF) and charging nodes (e.g., CRF and PDF) [Yan06]. Through the *Policy and Charging Control* (PCC) defined in 3GPP R7 [3GP07c], integration of QoS policy and charging rules can be realized in the IMS network to reduce the signaling costs. In the PCC architecture (Figure 9.6), the SBLP and the FBC functionality are utilized to integrate the QoS policy and charging control. In this architecture, the *Policy Enforcement Function* (PEF; Figure 9.6(b)) at the GGSN is responsible for QoS control of the IP service flows. According to the classification of a subscriber, the type of the application to be accessed by the subscriber, and the local control policy defined by the telecom operator, the IMS manages and controls the IP network resources (e.g., the allocated bandwidth) to the application and defines its priority. The process of IMS resource control is described as follows: During the setup of a session

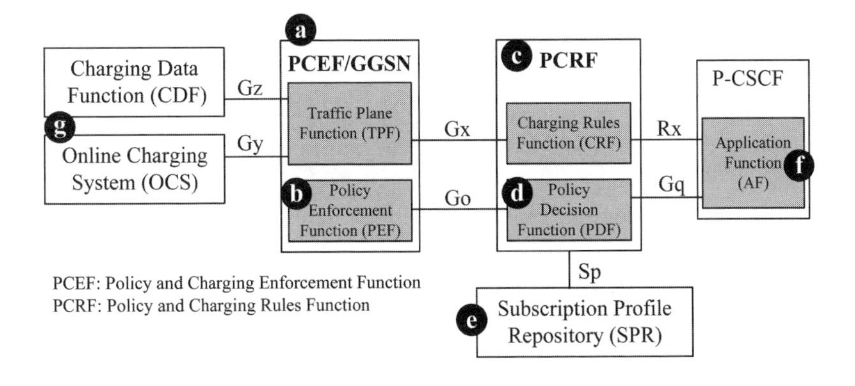

Figure 9.6 PCC integration architecture for IMS service

described in Section 7.1, the UE (MS) requests the network for related media parameters (such as codec, media type, and bandwidth) via the SDP contained in the SIP message. The AF/CSCF (Figure 9.6(f)) then forwards the SDP parameters to the PDF (Figure 9.6(d)) through the Gq interface [3GP07a]. The PDF controls the QoS policy by authorizing the related media parameters, according to the users' media messages and the local policy. After authorization, the authorized media parameters are returned to the UE and the resources for setting up the transmission bearer are reserved. The PDF then forwards the related IP QoS control parameter to the PEF/GGSN through the Go interface. As a device that executes the QoS control policy, the GGSN analyzes the source and destination IP addresses, and then controls and filters the IP flow.

The *Policy and Charging Enforcement Function* (PCEF; Figure 9.6(a)) is implemented in the GGSN. This function includes the TPF and the PEF. The *Policy and Charging Rules Function* (PCRF; Figure 9.6(c)) includes the CRF and the PDF (Figure 9.6(d)). The PCRF makes policy decision and provides IMS charging rules to the PCEF. The QoS decision is based on charging-related information and the service information. In this way, the charging rules are consistent with the QoS policy. According to the subscriber identity (e.g., IMSI), the PCRF can also request subscription information related to the bearer layer policies from the *Subscription Profile Repository* (SPR; see Figure 9.6(e)).

The AF hosts the service logic and communicates with the PCRF for the application level session information. This information indicates how to classify the service flows and to apply policy control at the GGSN. The offline and online charging systems (CDF and OCS; see Figure 9.6(g)) interact with PCEF for online credit control and the collection of offline charging information, respectively.

9.5 Concluding Remarks

In this chapter, we introduced the service data flow-based charging (FBC) architecture consisting of the *Charging Rule Function* (CRF), the *Traffic Plane Function* (TPF) and the *Application Function* (AF). The FBC allows breaking down of a PDP context into individual service data flows determined by the IP addresses and the port numbers. Therefore, the mobile operators can efficiently meter data usage for individual service flows. The FBC solution is used to accurately classify and charge the data packets by their content types through static and dynamic policies. For example, all packets addressed to a *Domain Name Server* (DNS) server are not accounted; all packets addressed to a service (identified by the IP address) are charged, and the IP-flows will appear in CDRs.

In the FBC architecture, the functionality of TPF/GGSN is enhanced to utilize IP packet filters such that different service flows within the same bearer session can be classified. This chapter gave GPRS and IMS service examples to explain how the FBC works. We also elaborated on the contents of the messages exchanged between the CRF and the AF, and between the CRF and the TPF.

At the end of this chapter, we briefly described the integration of *Policy and Charging Control* (PCC) for UMTS R7 and R8. In this architecture, the PCRF is designated for the determination of the policy rules in real time. With this node, a set of policy rules can be activated to verify access permission, check and debit credit balance etc., all in real time. The PCRF enforces these policy rules through its interaction with the PCEF, which handles the GPRS transport plane. More details of the PCC can be found in [3GP07c].

Review Questions

1. How does a telecom operator charge the transferred data for IMS signaling and IMS media with different tariff plans in UMTS Release 5? What are the disadvantages of this solution?
2. Describe the FBC architecture. Why is the AF only required for the IMS services?
3. Show how FBC works without PDP context.
4. What are the functionalities for the charging rule function and the traffic plane function?
5. Describe the messages for the Gx interface. What are the differences between them and the messages in the Rf interface?
6. Describe the messages for the Rx interface. What are the differences between them and the messages in the Gx interface?
7. Give an example of the charging rule for the IMS video call scenario.
8. Give an online IMS call example such that IMS signaling and the IMS media are transferred in two PDP contexts.
9. Describe the policy and charging integration architecture in UMTS Release 7. What are the advantages of this architecture?

References

[3GP05a] 3GPP, 3rd Generation Partnership Project; Technical Specification Group Services and Systems Aspects; Telecommunication management; Charging management; Charging data description for the Packet Switched (PS) domain (Release 5), 3G TS 32.215 version 5.9.0 (2005-06), 2005.

[3GP05b] 3GPP, 3rd Generation Partnership Project; Technical Specification Group Core Network and Terminals; Policy control over Go interface (Release 6), 3G TS 29.207 version 6.5.0 (2005-09), 2005.

[3GP05c] 3GPP, 3rd Generation Partnership Project; Technical Specification Group Services and System Aspects; End-to-end Quality of Service (QoS) concept and architecture (Release 5), 3G TS 23.207 version 5.10.0 (2005-09), 2005.

[3GP06a] 3GPP, 3rd Generation Partnership Project; Technical Specification Group Services and System Aspects; Overall high level functionality and architecture impacts of flow based charging; Stage 2 (Release 6), 3G TS 23.125 version 6.8.0 (2006-03), 2006.

[3GP06b] 3GPP, 3rd Generation Partnership Project; Technical Specification Group Core Network and Terminals; Charging rule provisioning over Gx interface (Release 6), 3G TS 29.210 version 6.7.0 (2006-12), 2006.

[3GP07a] 3GPP, 3rd Generation Partnership Project; Technical Specification Group Core Network and Terminals; Policy control over Gq interface (Release 6), 3G TS 29.209 version 6.7.0 (2007-06), 2007.

[3GP07b] 3GPP, 3rd Generation Partnership Project; Technical Specification Group Core Network and Terminals; Rx Interface and Rx/Gx signalling flows (Release 6), 3G TS 29.211 version 6.4.0 (2007-06), 2007.

[3GP07c] 3GPP, 3rd Generation Partnership Project; Technical Specification Group Services and System Aspects; Policy and charging control architecture (Release 7), 3G TS 23.203 version 7.5.0 (2007-12), 2007.

[Cou03] Courcoubetis C. and Weber, R., *Pricing Communication Networks: Economics, Technology, and Modelling*. John Wiley & Sons, Ltd., Chichester, UK, 2003.

[Fal00] Falkner, M., *et al.*, An overview of pricing concepts for broadband IP networks, *IEEE Communications Surveys & Tutorials*, 2nd Quarter: 2–13, 2000.

[IET05a] IETF, Diameter Network Access Server Application. IETF RFC 4005, 2005.

[IET05b] IETF, Diameter Credit-Control Application. IETF RFC 4006, 2005.

[Kou04] Koutsopoulou, M., *et al.*, Charging, accounting and billing management schemes in mobile telecommunication networks and the Internet, *IEEE Communications Surveys & Tutorials*, **6**(1): 50–58, 2004.

[Wimax] WiMAX Forum (www.wimaxforum.org)

[Yan06] Yan, T., Integration of policy and charging control in the IMS, *Communicate (Huawei)*, **25**: 41–43, 2006.

10

Billing for VoIP Services

The U.S. *Federal Communications Commission* (FCC) defines *Voice over Internet Protocol* (VoIP) as a technology that allows voice calls using a broadband access connection (e.g., ADSL or Cable Modem) instead of a regular (or analog) phone line. Some VoIP services only allow a user to call other people using the same VoIP service. Other VoIP services allow a user to call anyone who has a telephone number, including mobile and fixed-network numbers. Also, while some VoIP services only work over the user's computer or a special VoIP phone, other services allow the use of a traditional phone connected to a VoIP adapter.

The VoIP technology has significantly changed the billing model for telecommunications. Niklas Zennström stated, "The telephone is a 100-year-old technology. It's time for a change. Charging for phone calls is something you did last century." Indeed, a typical VoIP service provider offers free IP-to-IP calls. On the other hand, the users are charged for IP calls toward the PSTN, including mobile and fixed networks. Therefore, the charging mechanism is still required for VoIP services. This chapter uses commercial VoIP services offered by Artdio to illustrate VoIP *Call Detail Record* (CDR) generation. Then we use *call holding time* as an example to show how CDR information can be processed to provide useful telecommunications statistics. In particular, most billing plans are determined based on the call holding time statistics.

10.1 A VoIP Network Architecture

Figure 10.1 illustrates the network architecture of a *Session Initiation Protocol* (SIP)-based commercial VoIP system deployed by Artdio [Artd, Pan05]. This VoIP system consists of the following components:

Charging for Mobile All-IP Telecommunications Yi-Bing Lin and Sok-Ian Sou

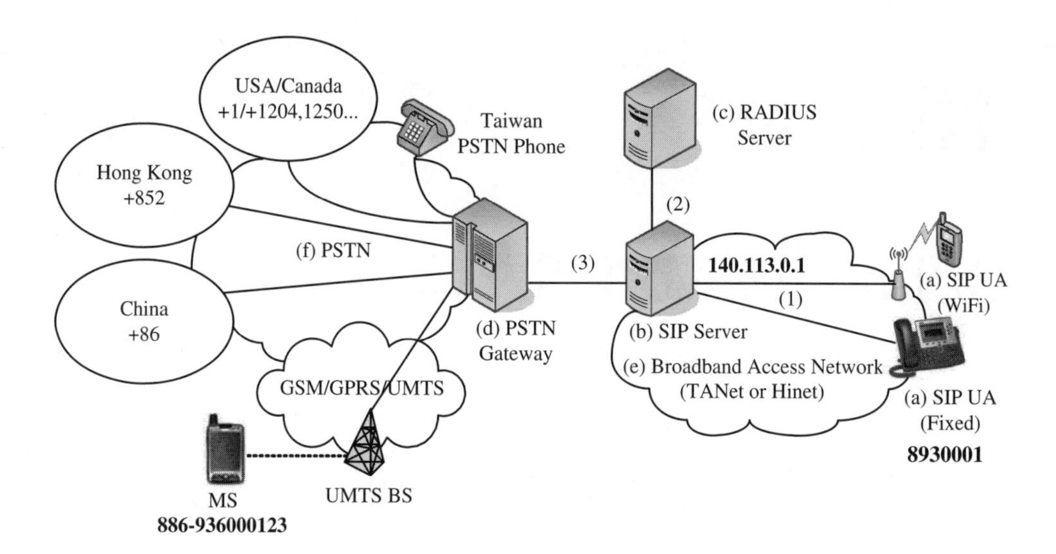

Figure 10.1 The Artdio VoIP network architecture

- A SIP *User Agent* (UA; see Figure 10.1(a)) can be a software-based VoIP phone (e.g., Windows Messenger or Artdio Spider UA in Figure 10.2(a)) or a hardware-based VoIP phone (e.g., Artdio hardphone illustrated in Figure 10.2(b)).
- The SIP server (Figure 10.1(b)), which provides SIP *registrar* and proxy functions, can support 50,000 subscribers with 500 concurrent *Real-time Transport Protocol* (RTP) connections [IET96].
- The *Remote Authentication Dial-In User Service* (RADIUS) server (Figure 10.1(c)) provides *Authentication, Authorization and Accounting* (AAA) for VoIP users. Call-related information such as *Call Holding Time* (CHT) is maintained in the database of this server.
- The PSTN Gateway (Figure 10.1(d); see also Figure 10.2(c)) provides VoIP–PSTN interworking between the broadband access network (Figure 10.1(e)) and the PSTN (Figure 10.1(f)). The PSTN gateway is a CISCO router with several E1 interfaces. Each interface supports 30 concurrent calls.

The SIP server, the RADIUS server and the PSTN gateway are located at the VoIP center in Taipei city. The SIP UAs connect to the SIP server through the broadband access network connections provided by domestic *Internet Service Providers* (ISPs) such as *Taiwan Academic Network* (TANet) or Hinet (Figure 10.1(1)). The total bandwidth between the VoIP center and other domestic ISPs is 18.14 Gbps. The SIP server connects to the RADIUS server and the PSTN gateway through 100Mbps Ethernet lines (Figure 10.1(2) and (3)). With SIP registration function, this VoIP system supports nomadic VoIP service. That is, the subscribers can move to any place in the world, and directly attach to the Artdio system for VoIP service through the Internet connectivity.

(a) **VoIP Softphone**

(b) VoIP Hardphone (Model IPF-2002L)

(c) VoIP Gateway (Model IPS-1000)

Figure 10.2 Examples of SIP user agents and VoIP gateway (reproduced by permission of ArtDio Co. Ltd)

A numbering plan similar to that for Taiwan's mobile phone service is exercised. For instance, VoIP users can dial 0936000123 to reach a FarEasTone (a Taiwan mobile operator) mobile phone user. Based on the numbering plan, the destinations of the calls can be distinguished and classified into several call types. Although VoIP is an IP-based application, its call holding time distribution does not follow those for other data applications that are typically modeled by the Pareto distribution [Yan06]. Its behavior does not exactly match PSTN-based voice service either. By investigating the CHT statistics derived from the Artdio CDRs, we provide useful insight to VoIP call holding times.

10.2 Call Detail Record Generation

For the VoIP system in Figure 10.1, the CDR information is collected by using the RADIUS protocol described in Chapter 5. The SIP server acts as a RADIUS client that interacts with the RADIUS server to create a CDR for each VoIP call. A VoIP CDR typically includes the following parameters:

- The *radacctid* parameter specifies a unique record identifier (e.g., 10243), which is automatically generated by the RADIUS server.
- The *username* parameter specifies the calling party (e.g., 8930001 in Figure 10.1(a)).
- The *nasipaddress* parameter indicates the IP address of the node that requests authentication (i.e., 140.113.0.1 for the SIP server in Figure 10.1).
- The *acctstarttime* and the *acctstoptime* parameters (e.g., 2008-01-22 09:01:32 and 2008-01-22 09:02:44, respectively) store the start time and the stop time of the call.
- The *acctsessiontime* parameter stores the call holding time (e.g., 00:01:12).
- The *calledstationid* and the *callingstationid* parameters record the telephone numbers of the called party (e.g., 886-936000123 for a mobile phone number in Taiwan; see Figure 10.1) and the calling party (i.e., 8930001), respectively.
- The *acctterminatecause* parameter indicates the reason why the VoIP call is terminated (e.g., User-Request).

Figure 10.3 illustrates the call setup and call release procedures where the calling party is a VoIP user and the called party is a mobile user. In the broadband access

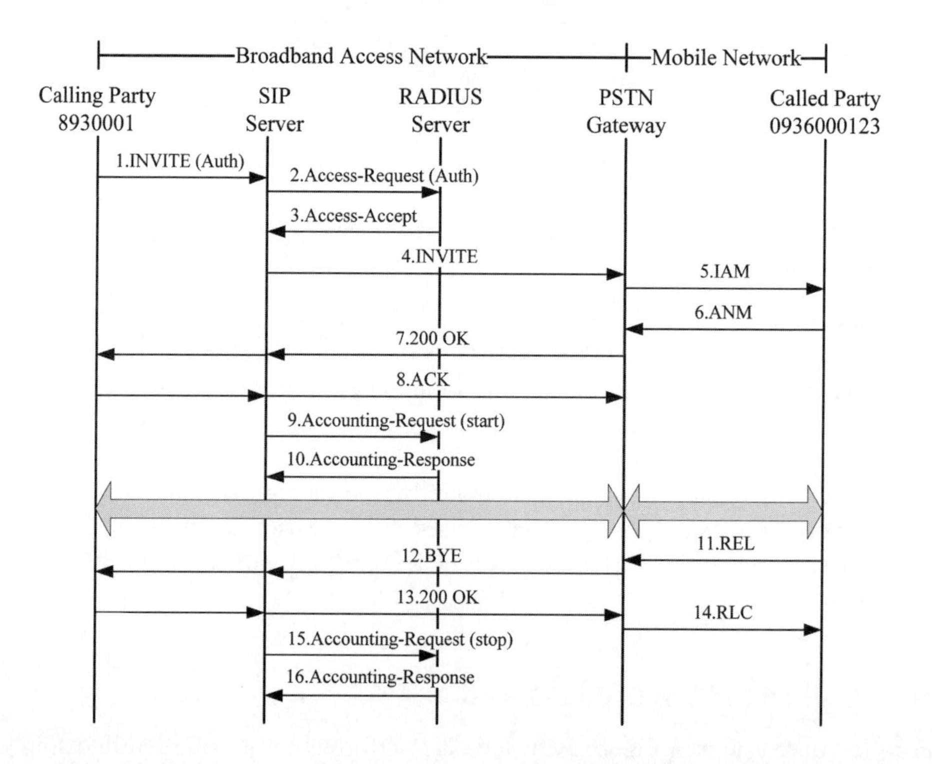

Figure 10.3 Message flow for call setup and call release from a broadband access network to the mobile network

network, call control is achieved by using SIP. On the other hand, the PSTN/mobile uses SS7 for call control as described in Chapter 2. The PSTN gateway is responsible for protocol translation between SIP and SS7 (i.e., it plays the role as a T-SGW and MGW in IMS; see Figure 2.11(k) and (c)). The calling party dials the local GSM number 0936000123 (or alternatively, +886936000123).

Step 1. The calling party sends an INVITE message to the SIP server. This message includes a *Request-URI* header field designating to the called party, a *From* header field indicates the calling party (8930001), a *To* header field indicates the mobile called party (0936000123) and several authentication parameters (e.g., username, realm and nonce).

Step 2. Upon receipt of the INVITE message, the SIP server acts as a RADIUS client and sends an Access-Request message to the RADIUS server. The Access-Request message contains the authentication parameters.

Step 3. The RADIUS server retrieves the user's record from its database by using the authentication parameters. If the authentication is successful, it replies with an Access-Accept message to the SIP server for authorizing the SIP request. At this point, the *username* parameter is confirmed and will be filled in the CDR at Step 10.

Step 4. Upon receipt of the Access-Accept message, the SIP server checks the *Request-URI*. Since 0936000123 is a mobile phone number, the SIP server forwards the INVITE message to the PSTN gateway.

Step 5. The PSTN gateway translates the INVITE message into an SS7 Initial Address Message (IAM) and sends it to the PSTN.

Step 6. After the called party has picked up the call, the PSTN replies with an Answer Message (ANM) to the PSTN gateway.

Step 7. The PSTN gateway translates this SS7 message into a SIP final response message 200 OK. This message is routed to the calling party following the reversed path of the INVITE message. That is, this message is sent to the SIP server and then forwarded to the calling party.

Step 8. The calling party replies with an ACK message to the PSTN gateway to confirm the receipt of the 200 OK message. The ACK message is sent to the PSTN gateway through the SIP server.

Step 9. Upon receipt of the ACK message, the SIP server sends an Accounting-Request message with status "start" to the RADIUS server to create a new CDR for this mobile VoIP call. This Accounting-Request message includes the *username* (8930001), the *nasipaddress* (140.113.0.1), the *acct-starttime* (2008-01-22 09:01:32), the *calledstationid* (886-936000123) and the *callingstationid* (8930001) parameters. In the created CDR, the *username* parameter is filled at Step 3, the *nasipaddress* parameter is the IP address of the SIP server, the *acctstarttime* parameter is retrieved from the local timer and the *callingstationed* parameter is retrieved from the *From* header field of the Accounting-Request message. Note that the *calledstationid* parameter is retrieved from the *To* header field of the message and is inserted a prefix 886 (for Taiwan) to indicate the call type (i.e., a Taiwan-mobile call in our example).

Step 10. Upon receipt of the Accounting-Request message, the RADIUS server generates a new CDR in its database, fills the above parameters into the CDR and acknowledges the request message by using an Accounting-Response message containing the *radacctid* parameter (10243) to identify the CDR record.

Step 11. Suppose that the called party hangs up the call when the conversation is complete. The PSTN sends the PSTN gateway a Release (REL) message to terminate the call.

Step 12. The PSTN gateway then generates a BYE message and sends it to the calling party through the SIP server.

Step 13. After the calling party has terminated the call, it sends a 200 OK message to the PSTN gateway through the SIP server. The 200 OK message indicates that the call is successfully terminated at the calling party.

Step 14. The PSTN gateway generates a Release Complete (RLC) message to the called party and releases the resources reserved for this call.

Step 15. Upon receipt of the 200 OK message, the SIP server issues an Accounting-Request message with status "stop" to the RADIUS server. This message includes the *radacctid* parameter (10243) received at Step 10, the *acctstop-time* parameter (2008-01-22 09:02:44) retrieved from the local timer and the *acctterminatecause* parameter (User-Request) that indicates the call is terminated by the user.

Step 16. The RADIUS server calculates the *acctsessiontime* parameter (00:01:12) and fills the parameters into the CDR (which is retrieved from the database based on the *radacctid* parameter). Finally, the RADIUS server replies with an Accounting-Response message to the SIP server.

From the above description, it is clear that after call setup, all CDR parameters for this call are stored in the database of the RADIUS server. When the call is released, the CDR is considered closed, and will be used for the billing and other purposes.

Table 10.1 The rate plan from Artdio VoIP to various destinations

Call destination	Prefix*	Rate (USD)
USA/Canada	+1, +1204, +1250...	0.021/minute
Hong Kong	+852	0.024/minute
China	+86	0.027/minute
Taiwan (fixed)	+886-[1~8]	0.027/minute
Taiwan (mobile)	+886-9	0.0097/6 seconds

* The prefix field includes the country code plus the area code or the service code.

Table 10.1 shows the charging rates for different call types in the Artdio VoIP system. In this table, the call destinations are distinguished by the prefix of the *calledstationid* parameter in the CDR. The CDR information can be utilized for collecting useful telecommunications statistics. We will show how to derive the CHT distributions based on the *acctsessiontime* parameter of the CDRs.

10.3 Deriving Call Holding Time Distributions

Based on the work in [Che07], we utilize the *hazard rate* (also known as the hazard function) [Ros96] to select a distribution for CHT. Then we use the *Kolmogorov–Smirnov* (K–S) test to validate if the selected distribution appropriately fits the measured data. **For readers who are not interested in statistic methods, this section can be skipped.**

In [Bar98, Bar00], the K–S test was utilized to select a distribution and its parameters to fit the measured data through the following steps:

Step 1. Select several candidate distributions such as Erlang, Normal and Exponential.

Step 2. For each of the selected distributions, apply the *Maximum Likelihood Estimation* (MLE) [Cas90] on the measured CHT data to estimate the parameters of the distribution.

Step 3. For each distribution with parameters determined in Step 2, use the measured data to calculate the distance D and the significance (confidence) level s of the K–S test, where

$$D = \varepsilon\left(\sqrt{n} + 0.12 + \frac{0.11}{\sqrt{n}}\right)$$

and

$$s = 2\sum_{i=1}^{\infty}(-1)^{i-1}e^{-2i^2 n\varepsilon^2}$$

In the above D and the s equations, n is the number of the measured data and ε is the maximum difference between the *Cumulative Distribution Function* (CDF) of the derived distribution and the measured data.

Step 4. Select the distribution with the smallest D and the largest s. This distribution is considered as the best fit to the CHT data.

The above procedure incurs two problems.

(a) The K–S test only applies to the examination of the following three cases [Dur75, Kie59, Smi48]:
   (i) whether the data set comes from a specific distribution where all parameters are known;

(ii) whether the data set comes from an Exponential distribution with unknown mean parameter; or

(iii) whether two data sets come from the same unknown distribution.

The above procedure used in [Bar98, Bar00] inspected whether the data set comes from a distribution with unknown parameters, which does not belong to any of the above three cases. Therefore, the K–S test cannot tell if the data set comes from the distribution selected at Step 4.

(b) The D and the s values are not suitable to identify whether the measured data come from a distribution with unknown parameters. In the real world, the data cannot be a "true" specific distribution such as Normal or Exponential. That is, the ε value in the D equation of Step 3 cannot be exactly zero. In the D and the s equations, more measured data (i.e., a larger n) result in larger D and smaller s. Therefore, it is difficult to set the "threshold" values of D and s to determine if the measured data comes from the distribution under test.

Based on the above discussion, it is not appropriate to directly use the K–S test to determine if the measured data come from the proposed distribution with unknown parameters. Instead, the K–S distance should be used to validate if the "already obtained" distribution can fit the measured data well after the distribution and its parameters are determined. We will show how to select a candidate distribution using the hazard rate, and then verify that the derived distribution fits the measured data by using the K–S test. We derive the CHT distribution for the VoIP calls to Taiwan's mobile networks (referred to as *Taiwan-mobile*). Other call types include calls to Taiwan's fixed network (referred to as *Taiwan-fixed*), to the USA/Canada, and to Hong Kong and China. The CHT distributions for these call types can be derived through the same procedure and the details are omitted.

For a probability distribution with the *Probability Density Function* (PDF) $f(t)$ and the CDF $F(t)$, the hazard function is defined as $h(t) = f(t)/(1 - F(t))$.

Figure 10.4 illustrates the histogram and the hazard rate for 20,000 measured CHT samples. The histogram in Figure 10.4(a) is similar to the PDF of a Gamma distribution or that of a Lognormal distribution. Figure 10.4(b) shows the hazard rate of the measured data, which does not fit the Gamma distribution. From the hazard rate curve in Figure 10.4(b), we are not able to identify its distribution. Therefore, we explore more characteristics of the measured CHTs by investigating the logarithm of the measured data (because the measured data has a long tail). Figures 10.4(c) and (d) illustrate the histogram and the hazard rate of the logarithm of the measured data (referred to as the logarithm measured data). Figure 10.4(c) shows that the histogram of the logarithm measured data is similar to that of a Normal distribution. Figure 10.4(d) shows that the hazard rate is an increasing function, which fits the hazard function of a Normal distribution. Based on the above observations, we confirm that the logarithm measured data match a Normal distribution, and therefore,

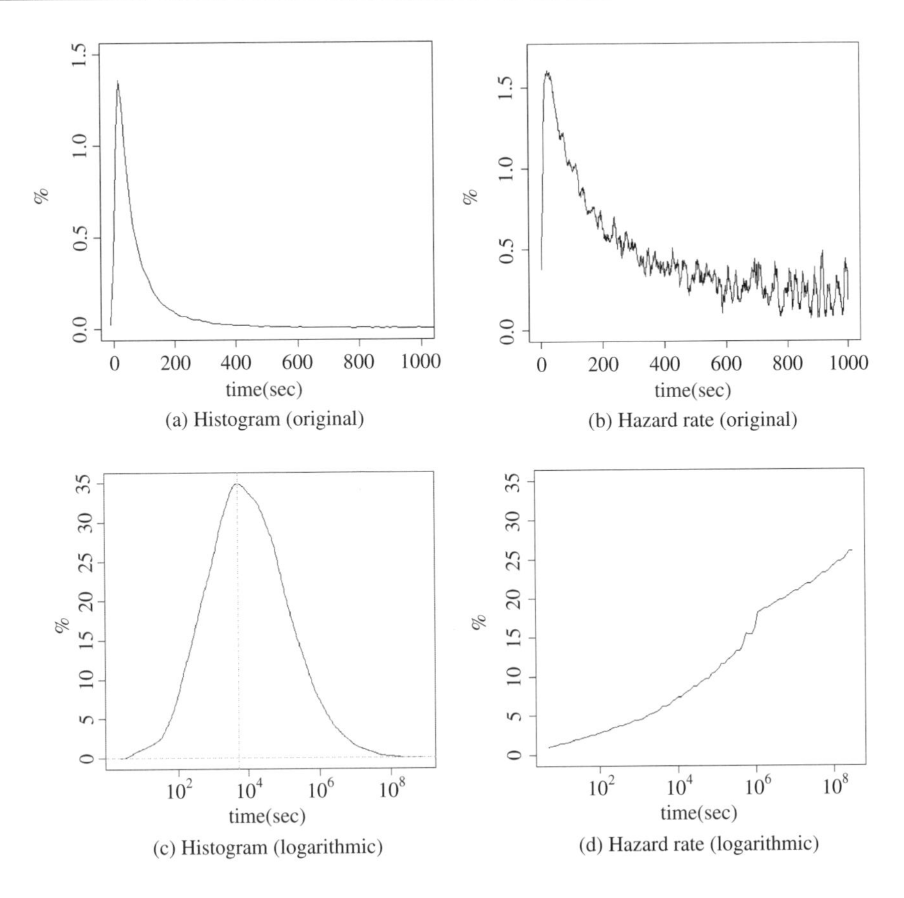

Figure 10.4 The CHT histogram and the hazard rate for mobile VoIP calls in Taiwan

the measured data match a Lognormal distribution. In addition, we observe that the curve in Figure 10.4(c) contains an asymmetric peak, which implies that this curve may be mixed by two Normal distributions. The mixed distribution has the following PDF

$$f(x) = \frac{p_1}{\sqrt{2\pi}\sigma_1} e^{-\frac{(x-\mu_1)^2}{2\sigma_1^2}} + \frac{p_2}{\sqrt{2\pi}\sigma_2} e^{-\frac{(x-\mu_2)^2}{2\sigma_2^2}} \quad (\text{where } p_2 = 1 - p_1)$$

To determine the parameter values for this mixed distribution, we use the *Expectation-Maximization* (EM) algorithm to find the *Maximum Likelihood Estimation* (MLE) [Dem97, Tan96]. By using the statistic tool R [Rpro], we obtain the parameters $\mu_1 = 3.606694$, $\sigma_1 = 0.9816$, $\mu_2 = 4.617415$, $\sigma_2 = 1.150835$, $p_1 = 0.6418951$, and $p_2 = 0.3581049$. Therefore, the CHT distribution has a mixed Lognormal PDF

$$f(x) = \frac{0.6418951}{x\sqrt{1.9270788\pi}} e^{\frac{(\ln x - 3.606694)^2}{1.9270788}} + \frac{0.3581049}{x\sqrt{2.648844\pi}} e^{-\frac{(\ln x - 4.617415)^2}{2.648844}}$$

To verify the derived distribution, we plot the histogram for the measured data (i.e., the bars) and the derived Lognormal distribution (i.e., the curve) in Figure 10.5. Through the K–S test, the derived distribution has the K–S distance D=0.00325696. Since the K–S distance is short, it indicates that the derived distribution provides good approximations to the measured data.

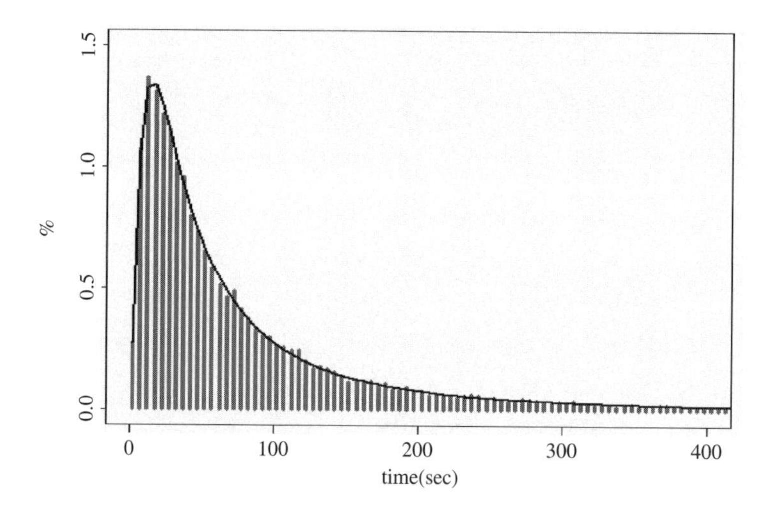

Figure 10.5 The histogram of the measured CHTs of mobile VoIP calls in Taiwan (the bars) and its fitting probability density function (the curve)

By applying the same procedure described above, we obtain the parameters of the derived mixed Lognormal distributions for other call types (see Table 10.2). The CHT distributions for Taiwan-mobile, *China-mixed* (mixing of mobile and fixed calls to China) and *Hong Kong-mixed* (mixing of mobile and fixed calls to Hong Kong) are mixed of two Lognormal distributions. The CHT distributions for Taiwan-fixed and *USA/Canada-mixed* (mixing of mobile and fixed calls to the USA/Canada) are mixed of three Lognormal distributions. Among the 120,000 CHT samples investigated, the proportions of Taiwan-mobile, Taiwan-fixed, Hong Kong-mixed, China-mixed, and USA/Canada-mixed are 17%, 30%, 2%, 49%, and 2%, respectively. In Table 10.2, the K–S distances show that the derived distributions provide good approximations to the measured data.

Based on the CHT statistics, the next section compares the characteristics among different call types.

Table 10.2 Parameters of the derived Lognormal distributions

Call type	Taiwan-mobile	Taiwan-fixed	Hong Kong-mixed	China-mixed	USA/Canada-mixed
K–S distance	0.0033	0.0063	0.0101	0.0226	0.0117
p_1	0.6419	0.0349	0.1116	0.0359	0.1521
p_2	0.3581	0.4665	0.8884	0.9641	0.2640
p_3	N/A	0.4987	N/A	N/A	0.5839
μ_1	3.6067	1.2065	2.3810	2.5196	1.4795
μ_2	4.6174	3.3832	4.5534	4.5124	3.0626
μ_3	N/A	4.4937	N/A	N/A	5.3206
σ_1	0.9816	0.2906	0.7947	0.2642	0.4456
σ_2	1.1508	0.7276	0.9593	1.2355	0.9016
σ_3	N/A	1.1538	N/A	N/A	1.2993
Mean (sec)	110	107	138	185	306
Coefficient of variance	2.07	2.16	1.47	1.89	2.83

10.4 Observations from the Call Holding Time Statistics

Previous studies indicated that the mean CHT for non-VoIP fixed phone calls is about 3 minutes [Chu99] or 113 seconds [Bol94]. Studies on non-VoIP mobile phone calls [Bar98, Bar00] indicated that the mean CHT is 40.6 seconds with *coefficient of variance* (*cv*) of 1.7 during working hours, and 63.3 seconds with $cv = 2.91$ during non-working hours. Measured data from Taiwan's mobile operators indicate that the mean CHT is 45 seconds [Far02]. Table 10.2 lists the means and coefficient of variances of the VoIP CHT distributions. Our study indicates that the mean CHT of Taiwan-mobile is 110 seconds (with $cv = 2.07$), which is much longer than the expected CHT of the non-VoIP mobile calls [Bar98, Bar00, Far02] and is close to that of the non-VoIP fixed phone calls [Bol94].

The mean CHT for all VoIP call samples measured in our study is 148 seconds, which is between the values measured in [Bol94] and [Chu99] for PSTN-based fixed network calls. Moreover, the coefficient of variances of all call types are larger than 1. This phenomenon indicates that the CHT distributions of all call types have high variations, and cannot be approximated by the Exponential distribution. We also note that most calls to Hong Kong and China are business calls, where the coefficient of variances are less than 2. Calls for other call types are mixed of business and residential calls, and the coefficient of variances are larger than 2. That is, the CHTs for business calls are more regular than that for residential calls. This phenomenon is also observed in [Bar98, Bar00].

It is interesting to note that in the investigated VoIP system, the farther the distances to the destinations, the longer the call holding times. This phenomenon is due to the fact that as compared with the charges for pure PSTN calls, more savings are expected for VoIP calls with farther distances. Also, the charges for VoIP calls to the USA/Canada are lower than those for other destinations (see Table 10.1), which is a major reason why CHTs for USA/Canada-mixed are longer than those for other destinations.

Figure 10.6(a) plots the PDFs for Taiwan-mobile and Taiwan-fixed. For CHTs of lengths ranging from 5 seconds to 10 seconds, and from 50 seconds to 400 seconds, there are more mobile calls than fixed calls. For other CHT values, more fixed calls are observed than mobile calls. In particular, there is a peak for fixed CHT PDF at 3–4 seconds, which means that a group of VoIP subscribers tend to make short calls to Taiwan-fixed. It is interesting to note that although the charge for Taiwan-mobile is 2.6 times higher than that for Taiwan-fixed (see Table 10.1), the average CHT for Taiwan-mobile is longer than that for Taiwan-fixed (see Table 10.2). As we explained before, this phenomenon is due to the fact that the charges of mobile calls for pure PSTN are much higher than that for mobile VoIP calls. Therefore, the users tend to

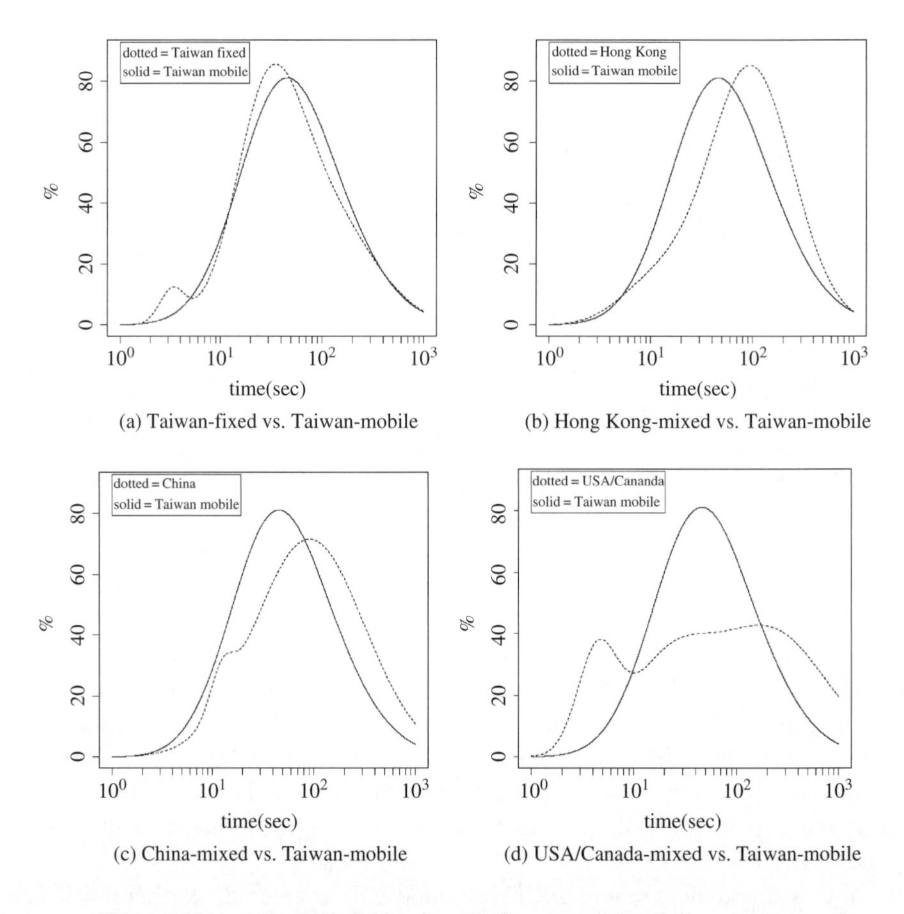

Figure 10.6 The probability density functions of various call types

make long mobile calls through the VoIP system. On the other hand, the charges of non-VoIP fixed calls are not much higher than that for VoIP calls, and the users tend to make long calls through pure PSTN for better voice quality.

Figure 10.6(b) indicates that for CHT lengths shorter than 70 seconds, more calls for Taiwan-mobile are observed than calls for Hong Kong-mixed. Figure 10.6(c) indicates that for CHT lengths shorter than 80 seconds, more calls for Taiwan-mobile are observed than calls for China-mixed. Figures 10.6(b) and (c) show that the CHT PDFs for Hong Kong and China are shifted to the right, which means that the CHT lengths for Hong Kong and China are longer than those for Taiwan-mobile. The variances of CHTs for Hong Kong-mixed and China-mixed are smaller than that for Taiwan-mobile. As we mentioned before, this phenomenon is due to the fact that most calls to Hong Kong and China are for business purposes. Therefore, these CHTs are more regular. Figure 10.6(d) indicates that for CHT lengths ranging from 10 seconds to 200 seconds, more calls for Taiwan-mobile are observed than calls for USA/Canada-mixed. The coefficient of variance statistics in Table 10.2 indicate that the calls to the USA/Canada are more irregular than the mobile calls within Taiwan.

10.5 Concluding Remarks

Voice over IP (VoIP) is one of the most popular applications in Internet. This chapter described the operation of Artdio commercial VoIP service to illustrate VoIP Call Detail Record (CDR) generation. Then we used call holding time as an example to show how CDR information can be processed to provide useful telecommunications statistics. The characteristics of call holding times for VoIP calls are quite different from traditional phone calls and are more close to those of fixed-network phone calls. The call holding time characteristics are userful in determining the billing plans. For example, a few years ago, a telecom operator changed its fixed-network billing plan from NT$1 per minute to NT$3 per 5 minutes. Although this telecom operator claimed that the users with long call holding times would benefit from the modified billing plan, the statistics from this operator obviously showed that most call holding times were shorter than 3 minutes.

Review Questions

1. Describe an abstract VoIP network architecture. Which node is responsible for interworking between this VoIP network and the PSTN? Which node is responsible for CDR generation?
2. Describe the parameters in a VoIP CDR. Are they the same as those in the PSTN CDR?
3. Draw the VoIP call setup procedure when the charging functions are implemented by using the Diameter protocol.

4. Based on Figure 10.3, can you describe the call setup procedure from the PSTN to a VoIP network? In Figure 10.3, can we execute Steps 15 and 16 before Step 13?
5. How are the hazard function and the K–S test used in deriving the call holding time distributions?
6. What is maximum likelihood estimation? How do you use it in deriving the call holding time distributions?
7. Our study indicates that the call holding time distributions are mixed Normal distributions. Can you find other studies that showed non-Normal call holding time distributions?
8. In Artdio VoIP system, we observed that the call holding times for international phone calls are longer than those for domestic phone calls. Explain why.
9. Based on the characteristics of call holding times in Figure 10.6, design the billing plans for Taiwan-mixed, Taiwan-mobile, Hong Kong-mixed, China-mixed and USA/Canada-mixed calls.

References

[Artd] Artdio Company Inc., http://www.artdio.com.tw.

[Bar98] Barcelo, F. and Jordan, J., Channel holding time distribution in cellular telephony, *Electronics Letters*, **34**(2): 146–147, 1998.

[Bar00] Barcelo, F. and Jordan, J., Channel holding time distribution in public telephony systems (PAMR and PCS), *IEEE Transaction on Vehicular Technology*, **49**(5): 1615–1625, 2000.

[Bol94] Bolotin, A.V., Modeling call holding time distributions for CCS network design and performance analysis, *IEEE Journal on Selected Areas in Communications*, **12**(3): 433–438, 1994.

[Cas90] Casella, G. and Berger, R.L., *Statistical Inference*. Wadsworth & Brooks/Cole, Belmont, CA, 1990.

[Che07] Chen, W.-E., Hung, H.-N. and Lin, Y.-B., Modeling VoIP call holding times for telecommunications, *IEEE Network*, **21**(6): 22–28, 2007.

[Chu99] Chung, M.Y., *et al.*, Performability analysis of common-channel signaling networks, based on signaling system 7, *IEEE Transactions on Reliability*, **48**(3): 224–233, 1999.

[Dem97] Dempster, A.P., Laird, N.M. and Rubin, D.B., Maximum likelihood from incomplete data via the EM algorithm, *Journal of the Royal Statistical Society*, Series B (Methodological), **39**(1): 1–38, 1997.

[Dur75] Durbin, J., Kolmogorov–Smirnov tests when parameters are estimated with applications to tests of exponentiality and tests on spacings, *Biometrika*, **62**(1): 5–22, 1975.

[Far02] Fareastone, *Private Communications*, 2002.

[IET96] IETF, RTP: A Transport Protocol for Real-Time Applications. IETF RFC 1889, 1996.

[Kie59] Kiefer, J., K-sample analogues of the Kolmogorov-Smirnov and Cramer-V. Mises tests, *Annals of Mathematical Statistics*, **30**(2): 420–447, 1959.

[Pan05] Pang, A.-C., *et al.*, A study on SIP session timer for wireless VoIP, *IEEE Wireless Communications & Networking Conference* (WCNC), **4**: 2306–2311, 2005.

[Ros96] Ross, S.M., *Stochastic Processes*. John Wiley & Sons Ltd., Chichester, UK, 1996.

[Rpro] The R Project for Statistical Computing, http://www.r-project.org/.

[Smi48] Smirnov, N., Table for estimating the goodness of fit of empirical distributions, *Annals of Mathematical Statistics*, **19**(2): 279–281, 1948.

[Tan96] Tanner, M., *Tools for Statistical Inference*. Springer Verlag, New York, 3rd edition, 1996.

[Yan06] Yang, S.-R., Dynamic power saving mechanism for 3G UMTS system, *ACM/Springer Mobile Networks and Applications* (MONET), online, 2006.

Appendix A

Connection Failure Detection for GTP'

In a telecommunications network, there must be an efficient way of handling the situation when the charging system does not respond within a timeout period, because the network node cannot wait indefinitely. This appendix shows how connection failure can be efficiently detected in GTP'. As described in Chapter 4, the GTP' protocol is utilized in the Ga interface to transfer the *Charging Data Records* (CDRs) from *GPRS Support Nodes* (GSNs) to *Charging Gateways* (CGs). For a UMTS PS service, the CDRs are generated based on the charging characteristics (data volume limit, duration limit and so on) of the subscription information for that service. Each GSN only sends the CDRs to the CG(s) in the same UMTS network. The CG maintains a *GSN list*. An entry in the list represents a GTP' connection to a GSN. This entry consists of pointers to the records in a non-volatile *CDR database*. Data stored in this database are analyzed and consolidated before the CG sends them to the billing system. The CG is associated with a *restart counter* that records the number of restarts performed at the CG. Details of this counter are given in Section 4.1.2. For the redundancy reason, a CG may also maintain a configurable list of peer CG addresses (e.g., to recommend other CGs to serve the GSNs).

A GSN maintains a list of CGs in the priority order (typically ranges from 1 to 100). This *CG list* can be configured by the *Operation and Management* (O&M) system. If a GSN unexpectedly loses its connection to the current CG, it may send the CDRs to the next CG in the priority list. An entry in the CG list describes the parameters for a GTP' connection, such as the pointers to the buffers containing the unacknowledged CDR packets. The entry also stores the restart counter of the corresponding CG.

After sending a GTP' request, a GSN may not receive a response from the CG due to network failure, network congestion or temporary node unavailability. In this case, 3GPP TS 29.060 [3GP06] defines a mechanism for request retry, where the GSN will

Charging for Mobile All-IP Telecommunications Yi-Bing Lin and Sok-Ian Sou

retransmit a message until either a response from the CG is received within a timeout period or the number of retries reaches a threshold value. In the latter case, the GSN–CG communication link is considered disconnected, and an alarm is sent to the O&M system. For a GSN–CG link failure, the O&M system may cancel CDR packets in the CG and unacknowledged sequence numbers (of the transient CDR packets) in the GSN.

Based on an analytic model proposed in [Hun06], this appendix studies the GTP' connection failure detection mechanism. Our study aims to provide guidelines for the mobile operators to select the parameters for GTP' connections between GSNs and CGs.

A.1 GTP' Failure Detection

This section describes the *path failure detection algorithm* that detects path failure between the GSN and the CG. Figure A.1 illustrates the data structures utilized by this algorithm.

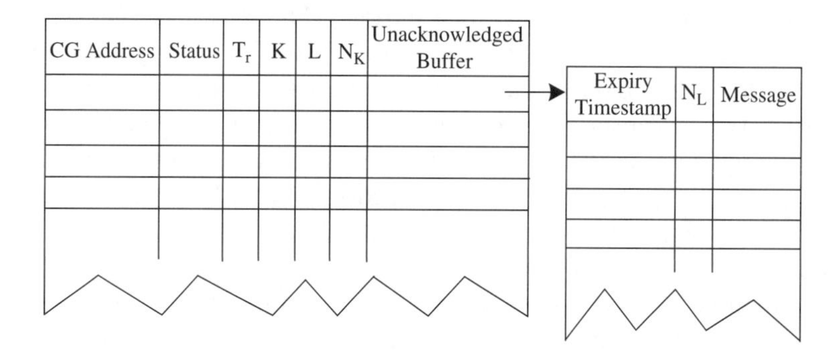

Figure A.1 Data structures for path failure detection algorithm

In a GSN, an entry in the CG list represents a GTP' connection to a CG. We describe the entry attributes related to the path failure detection algorithm as follows:

- The *CG address* attribute identifies the CG connected to the GSN.
- The *Status* attribute indicates if the connection is "active" or "inactive".
- The *Charging Packet Ack Wait Time* (T_r) defines the maximum elapsed time the GSN is allowed to wait for the acknowledgment of a charging packet; typical allowed values range from 1 millisecond to 65 seconds.
- The *Maximum Number of Charging Packet Tries* (L) defines the number of attempts (including the first attempt and the retries) the GSN is allowed to send a charging packet; typical L range is 1–16. When $L = 1$, there is no retry.

- The *Maximum Number of Unsuccessful Deliveries* (K) defines the maximum number of consecutive failed deliveries that are attempted before the GSN considers a connection failure occurs. Note that a *delivery* is considered failed if it has been attempted for L times without receiving any acknowledgment from the CG.
- The *Unsuccessful Delivery Counter* (N_K) attribute records the number of the consecutive failed delivery attempts.
- The *Unacknowledged Buffer* stores a copy of each GTP' message that has been sent to the CG but has not been acknowledged.

As shown in Figure A.1, a record in the unacknowledged buffer consists of an *Expiry Timestamp*, the *Charging Packet Try Counter* (N_L) and an unacknowledged GTP' message. The expiry timestamp is equal to T_r plus the time when the GTP' message was sent. This timestamp indicates the expired time of the message. The counter N_L records the number of the first attempt and retries that have been performed for this charging packet transmission. The path failure detection algorithm works as follows:

Step 1. After the connection setup procedure in Section 4.2 is complete, both N_L and N_K are set to 0, and the *Status* attribute is set to "active". At this point, the GSN can send GTP' messages to the CG.

Step 2. When a GTP' message is sent from the GSN to the CG at time t (Step 2 in Section 4.3), a copy of the message is stored in the unacknowledged buffer, where the expiry timestamp is set to $t + T_r$.

Step 3. If the GSN has received the acknowledgment from the CG before the expiry timestamp (Step 6 in Section 4.3), both N_L and N_K are set to 0. Return to Step 2.

Step 4. If the GSN has not received the acknowledgment from the CG after the expiry timestamp, N_L is incremented by 1. If $N_L = L$, then the charging packet delivery is considered failed. N_K is incremented by 1.

Step 5. If $N_K = K$, then the GTP' connection is considered failed. The *Status* attribute is set to "inactive". Otherwise, return to Step 2.

When the condition $N_K = K$ at Step 5 of the path failure detection algorithm is met, we consider that the path between the GSN and the CG is no longer available, and the GSN should be switched to another CG. However, besides link failure, unacknowledged packet transfers may also be caused by temporary network congestion. In this case, it is probably not desirable to perform CG switching (which is a very expensive operation). To avoid this kind of *"false" failure detection*, one may set large values for parameters T_r, L and K. However, large parameter values may result in delayed detection of "*true*" failures. Therefore, it is important to select appropriate parameter values so that true failures can be quickly detected while false failures can be avoided.

The GTP' mechanism can be evaluated by two output measures:

- P_d: the probability that a "good" GTP' connection is terminated due to false failure detection.
- $E[\tau_d]$: the expected detection delay of true failure; that is, the time between when the true failure occurs and when the failure is detected.

By investigating P_d and $E[\tau_d]$, Section A.2 uses numerical examples to show how to select appropriate input parameters for the path failure detection algorithm.

A.2 Numerical Examples

Based on the analytic model developed in [Hun06], this section shows how T_r, L and K affect the probability P_d of false failure detection and the expected delay $E[\tau_d]$ of true failure detection. We make the following assumptions:

- The round-trip transmission delay t_r between a GSN and a CG has the mean value $1/\mu$.
- The GTP' charging packet arrivals are a Poisson process with the rate λ_G.
- The expected lifetime of a GTP' connection (between when the GSN–CG link is established and when a true failure occurs) is $1/\lambda_f$.
- The Echo message arrivals are a deterministic stream with fixed interval T_e.

For the demonstration purposes, we consider $KL = 6$, $T_e = 18/\mu$ and the 2-Erlang distribution for the round-trip transmission delay t_r. We note that similar results are observed for different t_r distributions, and various KL and T_e values.

Effects of T_r and L on P_d. Figure A.2 plots P_d against T_r and the (K, L) pair, where $\lambda_G = \mu/18$, $KL = 6$ and $\lambda_f = 1 \times 10^{-5}\mu$. It is trivial that P_d is a decreasing function of T_r. A non-trivial result is that Figure A.2 quantitatively indicates how the T_r value affects P_d. When $T_r < 2/\mu$, increasing T_r significantly reduces P_d. On the other hand, when $T_r > 2/\mu$, increasing T_r does not improve the performance. Also, for small T_r, $L = 1$ outperforms other L setups. When T_r is large, the L (and thus K) values have same impact on P_d. Similar effects are observed for other λ_G values.

Effects of λ_f on P_d. Figure A.3 plots P_d as a function of T_r and λ_f, where $K = 6$, $L = 1$ and $\lambda_G = \mu/18$. This figure shows that P_d increases as λ_f decreases. When λ_f decreases (i.e., the system reliability improves but the transmission delay distribution remains the same as before), the GTP' connection lifetime becomes longer. Therefore, the opportunity for false failure detection increases. For $T_r = 1.6/\mu$, when the system reliability increases from $\lambda_f = 1 \times 10^{-5}\mu$ to $\lambda_f = 1 \times 10^{-6}\mu$, P_d increases by 2.72 times. This effect becomes insignificant when T_r is large (e.g., $T_r > 2.2/\mu$).

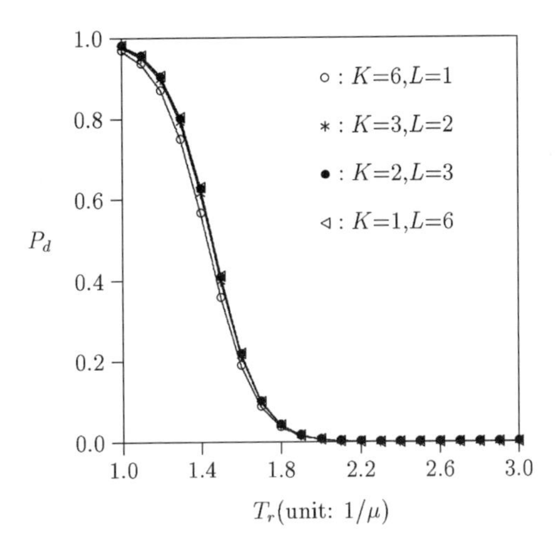

Figure A.2 Effects of T_r and L on P_d ($\lambda_G = \mu/18$, $\lambda_f = 1 \times 10^{-5}\mu$)

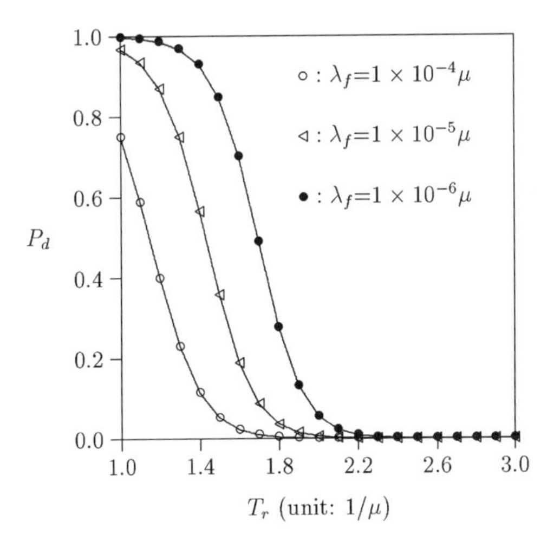

Figure A.3 Effects of T_r and λ_f on P_d($K = 6$, $L = 1$, $\lambda_G = \mu/18$)

Effects of λ_G on P_d. Figure A.4 plots P_d as a function of T_r and λ_G, where $K = 6$, $L = 1$ and $\lambda_f = 1 \times 10^{-5}\mu$. This figure shows that P_d increases as λ_G increases. When there are more GTP' message arrivals, it is more likely that false failure detection occurs. This effect is insignificant when T_r becomes large (e.g., $T_r > 2/\mu$).

Effects of λ_G on $E[\tau_d]$. Figure A.5 plots $E[\tau_d]$ as a function of T_r and λ_G, where $K = 6$ and $L = 1$. This figure shows that $E[\tau_d]$ significantly increases as λ_G decreases.

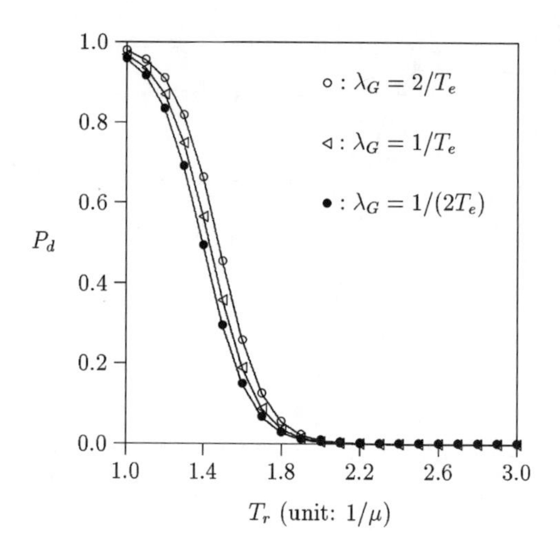

Figure A.4 Effects of T_r and λ_G on $P_d(K = 6,\ L = 1,\ \lambda_f = 1 \times 10^{-5}\mu)$

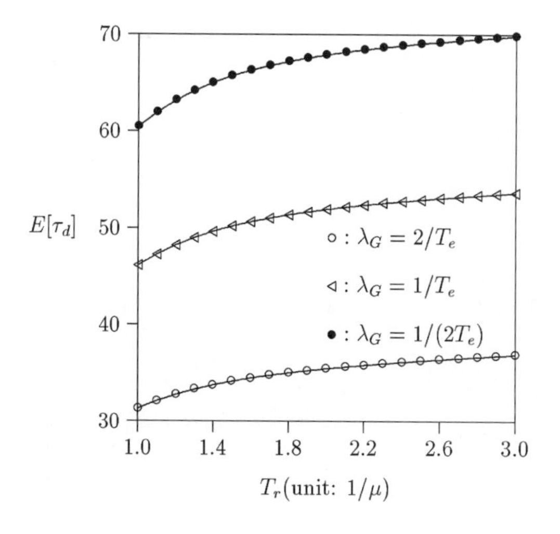

Figure A.5 Effects of T_r and λ_G on $E[\tau_d](K = 6,\ L = 1;\ E[\tau_d]$ is normalized by $1/\mu)$

Effects of (K, L) on $E[\tau_d]$. Figure A.6 plots $E[\tau_d]$ as functions of T_r and the (K, L) pair, where $\lambda_G = \mu$ and $\lambda_G = \mu/36$, respectively. These figures show that $E[\tau_d]$ is an increasing function of T_r, and $E[\tau_d]$ is more sensitive to the change of T_r when L is large than when L is small. When $\lambda_G = \mu$, $E[\tau_d]$ is larger for $L = 6$ than for $L = 1$. When $\lambda_G = \mu/36$, the opposite results are observed. On the other hand, when λ_G is small, the charging packets rarely occur in a short period, and it is likely that that $E[\tau_d]$ is smaller for $L = 6$ than for $L = 1$ in Figure A.6(b). Detailed explanation

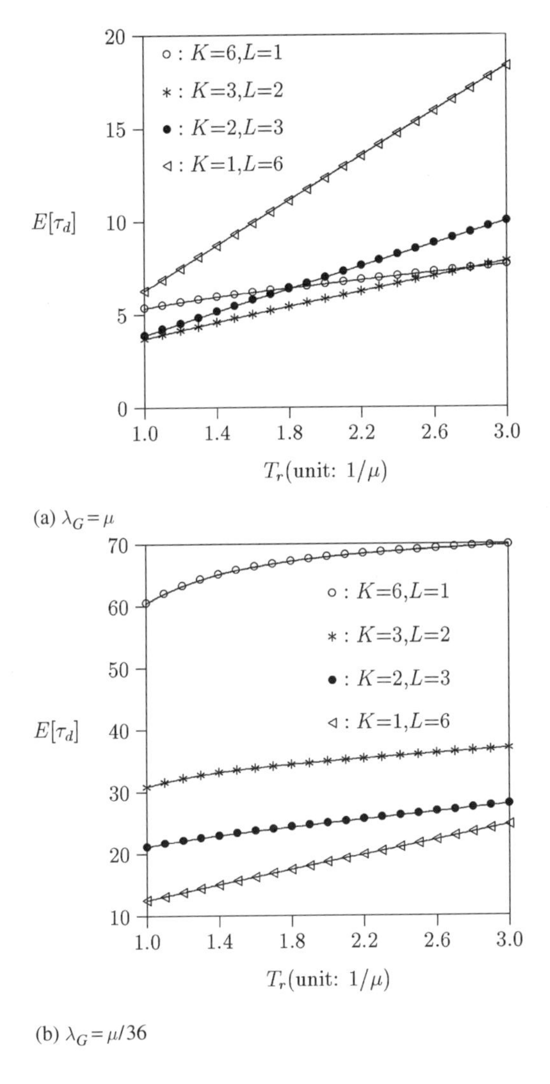

(a) $\lambda_G = \mu$

(b) $\lambda_G = \mu/36$

Figure A.6 Effects of T_r and (K, L) on $E[\tau_d]$ ($E[\tau_d]$ is normalized by $1/\mu$)

for these phenomena can be found in [Hun06], and the reader is encouraged to think about the intuitions behind these phenomena.

A.3 Concluding Remarks

In UMTS, the GTP' protocol is used to deliver the CDRs from GSNs to CGs. To ensure that the charging information is accurately delivered, availability for the charging system is essential. One of the most important issues on GTP' availability is connection failure detection. This appendix studied the GTP' connection failure detection mechanism specified in 3GPPTS 29.060 [3GP06] and 3GPPTS 32.215 [3GP05]. The output measures considered are the false failure detection probability P_d and the expected delay time $E[\tau_d]$ of true failure detection. We investigate how these two output measures are affected by input parameters including the Charging Packet Ack Wait Time T_r, the Maximum Number L of Charging Packet Tries and the Maximum Number K of Unsuccessful Deliveries. From this study, we make the following observations:

- When T_r is small, increasing T_r decreases P_d significantly. When T_r is sufficiently large, increasing T_r only has insignificant impact on P_d. On the other hand, increasing T_r always non-negligibly increases $E[\tau_d]$.
- Probability P_d increases as the charging packet arrival rate λ_G increases. This effect is insignificant when T_r becomes large.
- The effects of λ_G on $E[\tau_d]$ are not the same for different (K, L) setups. In our examples, when λ_G is large, $E[\tau_d]$ is larger for $L=6$ than for $L=1$. When λ_G is small, $E[\tau_d]$ is smaller for $L=6$ than for $L=1$. Therefore, the effects of λ_G should not be ignored when we select the L value.

In summary, a telecom operator should select the appropriate T_r, L and K values based on various traffic conditions. Also, the failure detection mechanism proposed in this appendix can be used in other protocols such as Diameter described in this book.

A.4 Notation

- K: the maximum number of consecutive failed deliveries that are attempted before the GSN considers a connection failure occurs
- L: the maximum number of attempts for a GTP' message that the GSN is allowed to send if it does not receive an acknowledgment
- N_K : the number of the consecutive failed delivery attempts
- N_L : the number of the first attempt and retries that have been performed for a charging packet transmission
- λ_G: the arrival rate of the GTP' charging packets
- $1/\lambda_f$: the expected lifetime of a GTP' connection
- $1/\mu$: the expected round-trip transmission delay for a GTP' message attempt
- P_d: the probability that a "good" GTP' connection is terminated due to false failure detection
- t_r: the round-trip transmission delay for a GTP' message attempt
- T_e: the fixed interval between two consecutive Echo messages
- T_r: the maximum elapsed time the GSN is allowed to wait for the acknowledgement of a GTP' message
- τ_d: the detection delay for a true failure

References

[3GP05] 3GPP, 3rd Generation Partnership Project; Technical Specification Group Services and System Aspects; Telecommunication Management; Charging Management; Charging data description for the Packet Switched (PS) domain (Release 5), 3G TS 32.215 version 5.9.0 (2005-06), 2005.

[3GP06] 3GPP, 3rd Generation Partnership Project; Technical Specification Group Core Network; General Packet Radio Service (GPRS); GPRS Tunneling Protocol (GTP) across the Gn and Gp Interface (Release 6), 3G TS 29.060 version 6.15.0 (2006-12), 2006.

[Hun06] Hung, H.-N., Lin, Y.-B., Peng, N.-F. and Sou, S.-I., Connection failure detection mechanism of UMTS charging protocol, *IEEE Transactions on Wireless Communications*, **5**(5): 1180–1186, 2006.

Appendix B

Charging for Integrated Prepaid VoIP and Messaging Services

As described in Section 2.4, the IMS nodes interact with each other through the *Session Initiation Protocol* (SIP) [3GP06a, IET02b]. With SIP, the IMS can effectively accommodate new applications. For example, with the growing demand of instant messaging services, RFC 3428 [IET02a] proposes the MESSAGE method, a SIP extension that allows transfer of instant messages over the Internet. With this extension, Internet services (such as mail and instant messaging [Rao03]) can be integrated with SMS/MMS through the IMS. The reader is referred to Sections 8.2.1 and 8.2.2 for charging examples of IMS messaging delivery using the MESSAGE method.

Billing mechanisms for messaging (including both SMS and MMS) and VoIP (especially for IMS calls toward the PSTN) are typically deployed for postpaid services. On the contrary, the prepaid billing mechanisms for combining these services are seldom studied in the literature. This appendix proposes a prepaid application server approach that can simultaneously process charging for SMS/MMS message deliveries and IMS calls for the same user account. This prepaid application server supports IMS calls toward the PSTN and messaging service toward UMTS. The notation used in this appendix is listed in Section B.5.

B.1 Prepaid Application Server of SIP-based Services

SIP-based prepaid services have been intensively investigated (see [Sou05, Vovida] and the references therein). However, how to manage and allocate credit for the SIP-based prepaid services is seldom mentioned in the literature. An appropriate solution is to deploy a new network entity to provide these functionalities without modifying the existing network entities (such as the OCS and the CSCF). This section proposes

Charging for Mobile All-IP Telecommunications Yi-Bing Lin and Sok-Ian Sou

a *Prepaid Application Server* (PAS) to manage credit allocation for both the prepaid SIP calls and messaging services in IMS. (In Appendices C, D and E, we will also consider credit reservation in various scenarios.)

Figure B.1 illustrates the PAS that interacts with several network nodes in the IMS network: The CSCF (Figure B.1(a)) utilizes SIP to provide control signaling for IP-based multimedia services. In Figure B.1(b), the MGCF controls the MGW to provide IP communications sessions. The T-SGW interworks the IMS with the PSTN. The *User Equipment* (UE; Figure B.1(c)) includes the SIP user agent to support IMS calls and SIP-based instant messaging services. The *IP-Message Gateway* (Figure B.1(d)) translates the SIP-based instant messages into short/multimedia messages. The *Short Message Service – Interworking Mobile Switching Center* (SMS-IWMSC; see Figure B.1(e)) sends/receives SMS to/from the IP-Message Gateway using standard SS7 *Mobile Application Part* (MAP) signaling [3GP06b], while the MMS Relay/Server (Figure B.1(f)) sends/receives MMS to/from the IP-Message Gateway. Note that the reader is referred to Figure 2.4 and Figure 2.9 for complete SMS and MMS architecture descriptions, respectively.

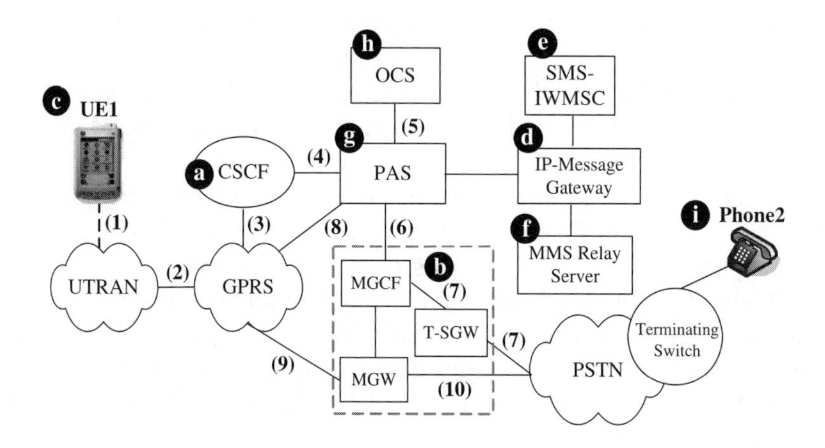

CSCF: Call Session Control Function
IWMSC: Interworking Mobile Switching Center
MGCF: Media Gateway Control Function
OCS: Online Charging System
PSTN: Public Switched Telephone Network
T-SGW: Transport Signaling Gateway
UTRAN: UMTS Terrestrial Radio Access Network
GPRS: General Packet Radio Service
MGW: Media Gateway
MMS: Multimedia Messaging Service
PAS: Prepaid Application Server
SMS: Short Message Service
UE: User Equipment

Figure B.1 The IMS environment for SIP-based prepaid services

By exchanging Diameter Credit Control Request (CCR) and Credit Control Answer (CCA) messages, the PAS (Figure B.1(g)) interacts with the OCS (Figure B.1(h)) to reserve the prepaid credit and process the online charging information. The PAS sets up prepaid IMS-to-PSTN calls through the MGCF, the MGW and the T-SGW, and supports prepaid SMS/MMS services through interaction with the IP-Message Gateway.

The prepaid call function in PAS works as follows: By utilizing the back-to-back user agent (B2BUA) technique [Vovida], the prepaid charging mechanism is inserted into a SIP connection session by breaking it into two sub-sessions. In this way, the PAS can monitor and terminate the call session when the prepaid credit is depleted. To set up an IMS-to-PSTN call for an IMS user UE1, the signaling message is first routed from UE1 to the CSCF. The CSCF identifies the charging type of the service (i.e., prepaid or postpaid). For a postpaid SIP request, the call is set up by standard IMS procedures described in Section 7.1. For a prepaid SIP request, the CSCF forwards the request to the PAS for further authorization and call session control. The details will be given in Section B.2.

B.2 Charging Integration for Prepaid Calls and Instant Messaging

This section first describes prepaid IMS-to-PSTN call setup and release. Then we describe instant message delivery through the prepaid mechanism. Finally, we describe the PAS charging policy for integrating IMS-to-PSTN calls and instant messages.

B.2.1 Prepaid IMS-to-PSTN Call Setup and Release

Consider the scenario where an IMS user (UE1; Figure B.1(c)) makes a VoIP call to a PSTN user (Phone2 in Figure B.1(i)). The call setup message flow is similar to that described in Sections 7.1 and 10.2. The major difference is that in this case, the PAS (i.e., the prepaid B2BUA) breaks the SIP session between UE1 and the MGCF into one sub-session (**subsession 1**) between UE1 and the PAS (Figure B.1(g)) and another sub-session (**subsession 2**) between the PAS and the MGCF (Figure B.1(b)).

In Figure B.1, the call setup request from UE1 (a prepaid user) is first routed to the PAS through the CSCF (path (1)→ (2)→ (3)→ (4)) to establish **subsession 1**. Then the PAS interacts with the OCS to reserve the subscriber's credit (path (5)). When the user's authorized time (i.e., the prepaid credit) is granted, the PAS generates a new call setup message for **subsession 2**, and then sends it to the MGCF through path (6). The MGCF instructs the T-SGW to deliver the call setup request to Phone2 via path (7). When the called party answers, the answer message is sent to the PAS through path (7)→ (6). Then the PAS responds with the accept message to UE1 through path (4)→ (3)→ (2)→ (1). Finally, the PAS starts an authorized session timer with the value (the available prepaid credit) granted from the OCS. UE1 then sends the acknowledgment message (for **subsession 1**) to the PAS through path (1)→ (2)→ (8); note that the acknowledgment message need not be routed through the CSCF. The PAS sends the acknowledgment message (for **subsession 2**) to the MGCF through path (6). At this point, the MGW opens a *Real-time Transport Protocol* (RTP) [IET96] connection so that UE1 can communicate with Phone2 through the MGW. The media path for the prepaid call is (1)→(2)→ (9)→ (10). If the prepaid credit is exhausted (i.e., the authorized session timer expires) before the conversation is complete, the call is forced to terminate by the PAS. When the authorized session timer expires, the PAS sends the call release messages to both UE1 (path (8)→ (2)→ (1)) and Phone2 through the MGCF (path (6)→ (7)). Specifically, the MGCF instructs the MGW to release the RTP connection, and the subsequent voice packets delivered between UE1 and Phone2 are not allowed to pass through the MGW. The MGCF instructs the T-SGW to ask Phone2 to release the call via path (7). Finally, the PAS triggers the OCS to debit the user account through path (5). The message flow for the prepaid call setup and force-termination is described below (see Figure B.2).

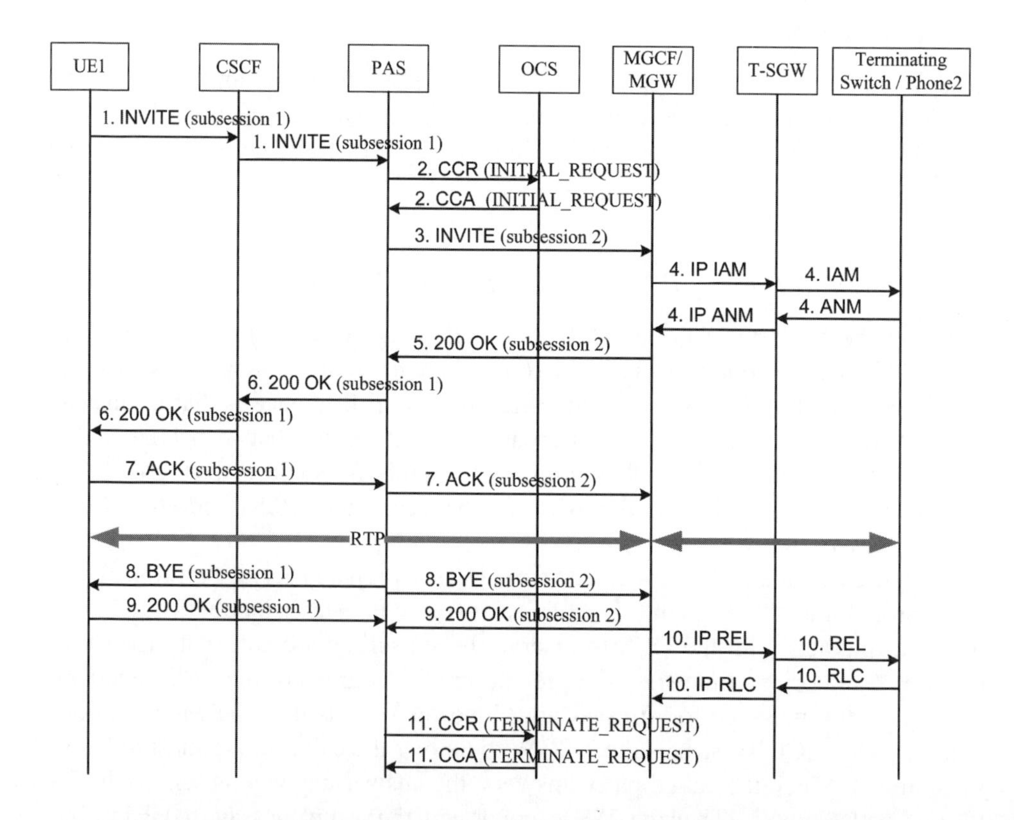

Figure B.2 Message flow for prepaid IMS-to-PSTN call setup and release

Step 1. The SIP INVITE request from UE1 is first routed to the PAS through the CSCF.

Step 2. The PAS sends a CCR message with *CC-Request-Type* "INITIAL_REQUEST" to the OCS to reserve the subscriber's credit. The OCS replies to the PAS with the CCA message containing the credit information (i.e., the user's authorized time).

Step 3. When the user's authorized time quota is granted, the PAS generates a new INVITE message for **subsession 2**, and then forwards the message to the MGCF.

Step 4. The MGCF translates the SIP INVITE message into IP-based Initial Address Message (IAM), and sends the IP-based IAM to the T-SGW. This message instructs the T-SGW to deliver the SS7 IAM (the SS7 call setup message [Lin01]) to the terminating switch of Phone2. When the called party answers, the SS7 Answer Message (ANM) is sent to the T-SGW. Then the T-SGW sends an IP-based ANM message to the MGCF.

Step 5. The MGCF instructs the MGW to allocate the data bandwidth according to the negotiated QoS profile specified in the SDP. Then the MGCF sends a final response 200 OK to the PAS.

Step 6. The PAS sends a final response 200 OK to UE1 to indicate the allocated QoS profile for this connection.

Step 7. If UE1 agrees to the negotiated QoS profile, it sends the ACK message (for **subsession 1**) to the PAS. The PAS sends the ACK message (for **subsession 2**) to the MGCF. At this point, the MGW opens an RTP connection so that UE1 and Phone2 can deliver voice packets through the MGW. Finally, the PAS starts an authorized session timer with the value (the available prepaid credit) granted from the OCS. The timer decrements the prepaid credit during the call session.

Step 8. If the prepaid credit is exhausted before the conversation is complete (i.e., the authorized session timer expires), the call is forced to terminate by the PAS. The PAS will send the BYE messages to both UE1 and the MGCF. Then the MGCF instructs the MGW to release the RTP connection, and the subsequent voice packets delivered between UE1 and Phone2 are not allowed to pass through the MGW.

Step 9. Both UE1 and the MGCF understand that the call is forced to terminate, and reply to the PAS with the 200 OK messages.

Step 10. The MGCF instructs the T-SGW to send the SS7 Release (REL) message to the terminating switch of Phone2. The terminating switch replies to the T-SGW with the SS7 Release Complete (RLC) to indicate that the resources in the PSTN (e.g., the trunk for voice) have been released. Then the T-SGW sends an IP-based RLC message to the MGCF.

Step 11. Finally, the PAS sends a CCR message to the OCS. The *CC-Request-Type* of the message is "TERMINATION_REQUEST". The OCS debits the user account, and then replies the PAS with a CCA message to complete the charging process.

B.2.2 Prepaid Instant Messaging Delivery

The message flow for prepaid instant messaging service is similar to the one described in Section 8.2.1 except for the introduction of the PAS. The details are shown in Figure B.3, and are described in the following steps:

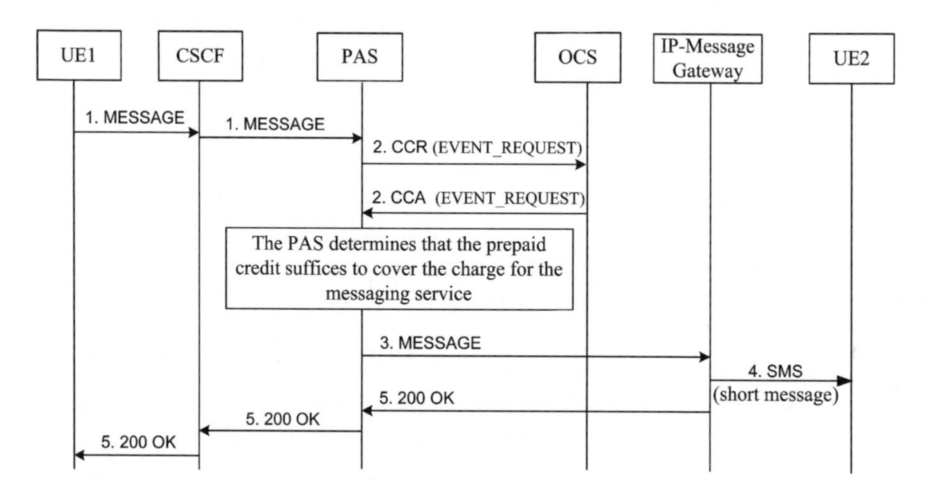

Figure B.3 Message flow for prepaid messaging delivery

Step 1. A prepaid user UE1 sends the SIP MESSAGE request to the PAS through the CSCF (including P-CSCF and S-CSCF). The instant message to be delivered is attached in this SIP request.

Step 2. The PAS sends a CCR message to the OCS. The *CC-Request-Type* is "EVENT_REQUEST". The OCS debits the subscriber's prepaid account, and then replies to the PAS with the CCA message containing the cost information (i.e., the amount of credit charged for this service).

Step 3. If the amount of prepaid credit in UE1's account is large enough to cover the charge for the messaging service, the PAS forwards the MESSAGE request to the IP-Message Gateway.

Note that if the messaging request arrives during a prepaid call, then the PAS already obtained all prepaid credit units from the OCS (Step 2 in Figure B.2) and Step 2 in Figure B.3 is skipped.

Step 4. The IP-Message Gateway retrieves the instant message content from the MESSAGE request and delivers it to UE2 through the SMS/MMS network described in Sections 2.2 and 2.3.2.

Step 5. The IP-Message Gateway sends the SIP 200 OK message to UE1 through the PAS and the CSCF.

Step 4 in the above message flow assumes that the SMS/MMS delivery is successful. The reader is encouraged to redraw the message flow when the instant message delivery fails. In this case, the OCS should reclaim the prepaid credit previously deducted for the instant message service.

B.2.3 Charging Policy of the Prepaid Application Server

The PAS for pure voice service or pure instant message service can be easily implemented as described in Subsections B.2.1 and B.2.2. However, when both prepaid voice and prepaid messaging are simultaneously offered, charging becomes complicated. That is, during a prepaid IMS-to-PSTN call session, the user may attempt to send an instant message. Deduction of prepaid credit for sending out this message may result in insufficient amount of credit left for the in-progress prepaid call, especially when the delivered instant message is a huge multimedia data file which causes large credit deduction. In this case, the IMS-to-PSTN call is forced to terminate. Therefore, a policy is required to determine if the prepaid instant message can be sent out without causing force-termination of an ongoing call. To avoid unnecessary force-termination, we propose a policy that sets a threshold X_T to reduce the possibility of force-termination for IMS-to-PSTN calls.

Consider the timing diagram in Figure B.4. A prepaid call arrives at t_1 and completes at t_3. Without loss of generality, we assume that each time unit of the call is charged for one credit unit. A prepaid authorized timer for this call starts at t_1 and expires at t_4. That

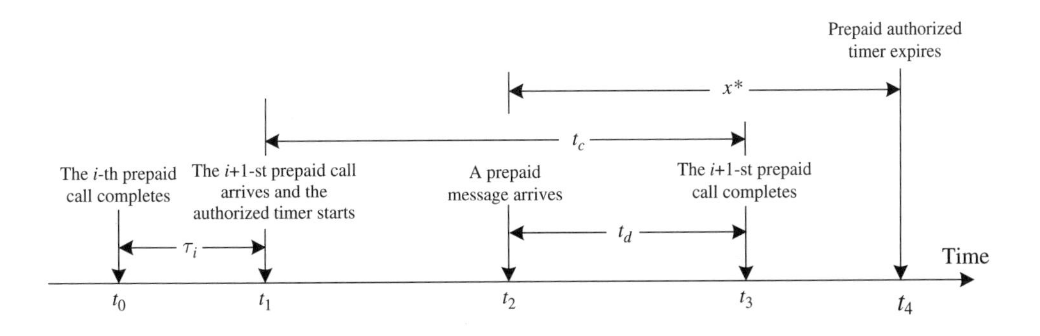

Figure B.4 Timing diagram for prepaid calls and messaging deliveries

is, at time $t_1$the remaining amount of the prepaid credit in the user's account is t_4–t_1. Upon receipt of the prepaid instant message request at t_2, where $t_1 < t_2 < t_3$, the PAS estimates if the remaining credit $x^* = t_4$–t_2 suffices to support both the in-progress call and the instant message service. Assume that the prepaid instant message service is charged for a fixed amount $T_m$of credit units. The policy used in the PAS is described as follows:

- If the remaining prepaid credit $x^* \geq X_T + T_m$, the instant message is sent immediately and the amount of remaining prepaid credit is reduced to x^*–T_m.
- If not (i.e., $x^* < X_T + T_m$), the instant message is stored. After the IMS-to-PSTN call is completed, the PAS checks if the amount of remaining credit is larger than T_m. If so, the instant message is delivered and the prepaid credit is deducted by the amount T_m. Otherwise, the instant message is aborted and the user is requested to refill the prepaid account. Consider an instant message service request that arrives at time t_2 in Figure B.4. If $x^* < X_T + T_m$, then the decision for delivering this instant message is postponed until t_3 (when the in-progress call is finished). In this case, the extra delayed time for instant message delivery is $t_d = t_3$–t_2.

We note that in the PAS, all credit units in a prepaid user's account are allocated to the in-progress IMS-to-PSTN call, and the credit allocation for instant messaging is controlled by the PAS. Therefore, the PAS does not need to interact with the OCS for the message service. In Appendices C and D, the charging system exercises a credit reservation model, and only assigns a fixed amount of credit to the in-progress call. When the assigned credit depletes, the charging system assigns extra credit to the call. The reader is encouraged to accommodate the credit reservation model in the PAS.

B.3 Performance for the PAS Charging Policy

Performance of the charging policy described in Section B.2 is significantly affected by the selection of the threshold X_T. If X_T is set too large, the PAS may reserve too many prepaid credit units, and therefore instant message deliveries are unnecessarily delayed. If X_T is set too small, the PAS may reserve too few prepaid credit units for the remaining prepaid call, and therefore results in unnecessary force-terminations (UFTs). In [Sou07], we proposed an analytic model to derive the UFT probability P_{UFT} and the expected extra delay $E[t_d]$ for instant message delivery. Let X be the amount of the initial prepaid credit. We make the following assumptions:

1. The prepaid instant message arrivals are a Poisson stream with rate λ_m.
2. Let τ_i be the interval between when the i-th prepaid call completes and when the i+1-st prepaid call starts (i >0). In Figure B.4, $\tau_i = t_1$–t_0. By convention, τ_0

is the interval between when the prepaid account is first activated (or recharged) and when the first prepaid call arrives. Interval τ_i is assumed to be exponentially distributed with the mean $1/\lambda$.

3. The prepaid call holding time t_c (i.e., $t_c = t_3 - t_1$ in Figure B.4) has a Gamma distribution with the mean $E[t_c]$ and the variance V_c.

Based on the analytic model and the simulation experiments in [Sou07], this section investigates the performance of the PAS. (The reader is encouraged to derive the output measures based on the above assumptions.) The input parameter X_T and the output measure $E[t_d]$ are normalized by the mean of the call holding time $E[t_c]$. For the demonstration purpose, we assume that the prepaid instant message service is charged for $T_m = 5$ credit units and the expected inter-call arrival time $1/\lambda = 50$ time units (and every time unit is charged for one credit unit). The expected inter-instant message arrival time $1/\lambda_m$ and the initial prepaid credit X are normalized by $E[t_c] = 4T_m$.

Effects of the threshold X_T. In Figures B.5– B.7, when X_T is small, increasing X_T reduces P_{UFT} significantly. When $X_T \geq 5E[t_c]$, the effect of X_T on P_{UFT} becomes insignificant. On the other hand, increasing X_T always increases $E[t_d]$. In the scenarios investigated in this appendix, it is appropriate to choose $X_T = 5E[t_c]$.

Effects of the variance V_c for the call holding time t_c. Figure B.5 plots P_{UFT} and $E[t_d]$ against X_T and the variance V_c of the call holding time t_c with Gamma distribution, where $X = 25E[t_c]$ and $\lambda_m = 2/E[t_c]$. Figure B.5 shows that when $X_T \geq 5E[t_c]$, both P_{UFT} and $E[t_d]$ increase as V_c increases. This phenomenon is explained as follows: As V_c increases, more long and short t_c periods are observed. The prepaid instant message (i.e., a random observer) is more likely to fall in the long t_c periods than the short t_c periods, and larger residual call holding times t_d are expected for larger variance V_c. Therefore, both the measures P_{UFT} and $E[t_d]$ degrade as V_c increases.

Effects of the expected inter-instant message arrival time $1/\lambda_m$. Figure B.6 plots P_{UFT} and $E[t_d]$ against X_T and λ_m, where $X = 25E[t_c]$. Figure B.6(a) shows that P_{UFT} decreases as $1/\lambda_m$ increases. When $1/\lambda_m$ decreases, more instant message deliveries are likely to occur during an in-progress call. Therefore the UFT probability P_{UFT} increases. For $X_T = 2E[t_c]$, when $1/\lambda_m$ decreases from $50E[t_c]$ to $5E[t_c]$ and from $5E[t_c]$ to $0.5E[t_c]$, P_{UFT} increases by 9.13 and 9.01 times, respectively. This effect becomes insignificant when X_T is large (e.g., $X_T \geq 6E[t_c]$). P_{UFT} is not significantly affected by $1/\lambda_m$ when $1/\lambda_m > 10E[t_c]$. Figure B.6(b) shows that $E[t_d]$ is insignificantly affected by $1/\lambda_m$. This phenomenon can be explained as follows: Since the prepaid instant messages are random observation points of prepaid call holding intervals, the delivery delays are not significantly affected by the instant message arrival rate λ_m.

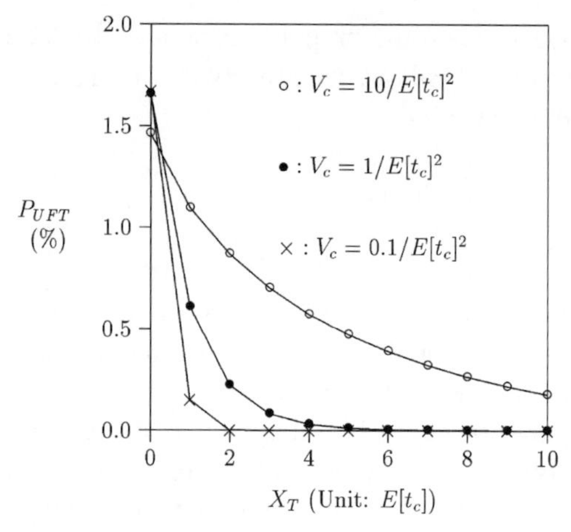

(a) Effect on P_{UFT}

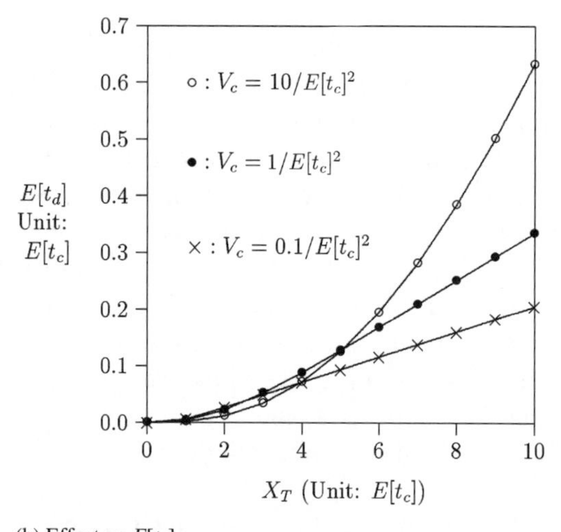

(b) Effect on $E[t_d]$

Figure B.5 Effect of V_c on P_{UFT} and $E[t_d](X = 25E[t_c])$ and $\lambda_m = 2/E[t_c]$)

Effects of the initial prepaid credit amount X. Figure B.7 plots P_{UFT} and $E[t_d]$ against X_T and X, where $\lambda_m = 2/E[t_c]$. This figure shows that both P_{UFT} and $E[t_d]$ decrease as X increases. When the initial prepaid credit X increases, there are more credit units left for a prepaid call, and it is more likely that there are enough credit units for both the remaining call and the instant message (i.e., the amount of the prepaid credits left is larger than $t_c + T_m$). Therefore, both output measures P_{UFT} and $E[t_d]$

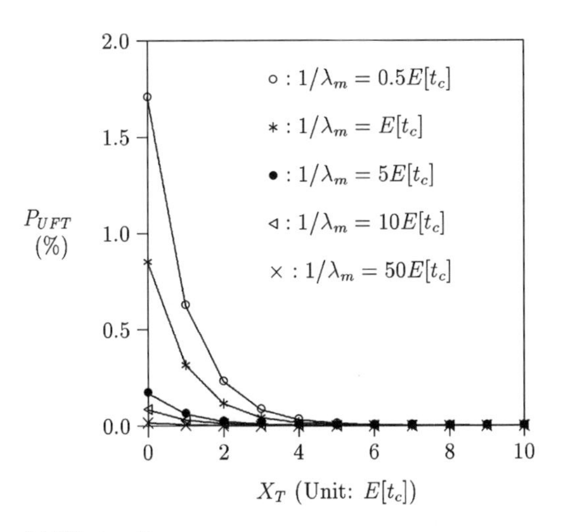

(a) Effect on P_{UFT}

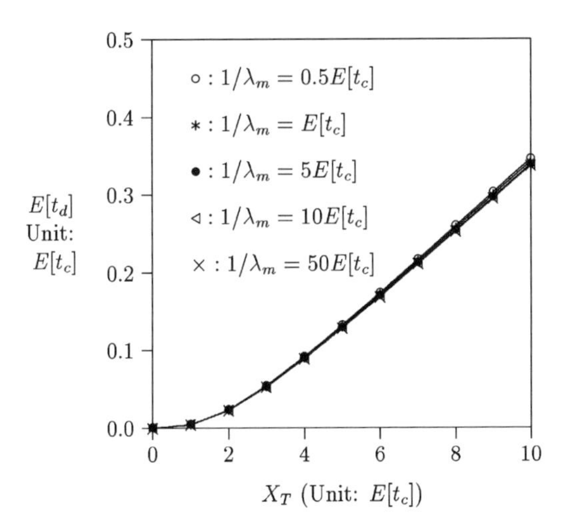

(b) Effect on $E[t_d]$

Figure B.6 Effect on $1/\lambda_m$ on P_{UFT} and $E[t_d](X = 25E[t_c])$

improve as X increases. For $X_T = 3E[t_c]$, when X increases from $20E[t_c]$ to $50E[t_c]$, P_{UFT} decreases by 58.64% and $E[t_d]$ decreases by 62.28%. When X increases from $50E[t_c]$ to $100E[t_c]$, P_{UFT} decreases by 49.96% and $E[t_d]$ decreases by 50.87%. When the initial prepaid credit can support more than 150 voice calls (i.e.,$X > 150\ E[t_c]$), increasing X only has insignificant impacts on P_{UFT} and $E[t_d]$.

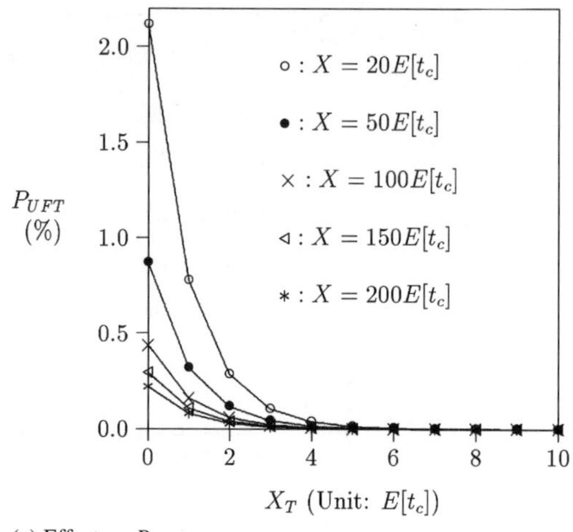

(a) Effect on P_{UFT}

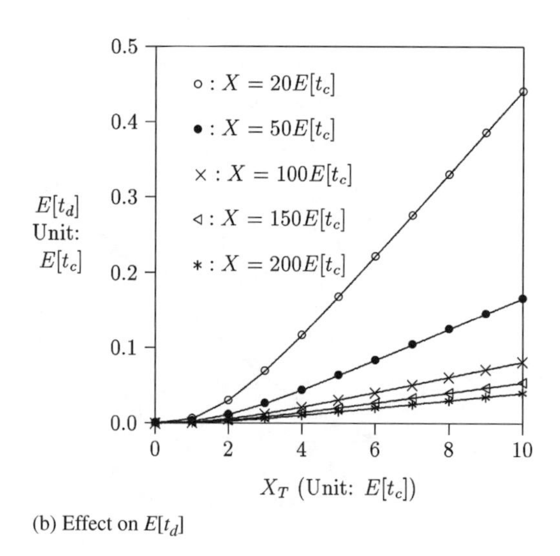

(b) Effect on $E[t_d]$

Figure B.7 Effect of X on P_{UFT} and $E[t_d](\lambda_m = 2/E[t_c])$

B.4 Concluding Remarks

This appendix proposed a SIP-based prepaid application server to handle both the prepaid IMS-to-PSTN calls and instant messaging services in UMTS. When both voice and messaging are simultaneously offered, a policy is required to determine if a prepaid instant message can be sent out during an in-progress call without force-terminating this call. A threshold X_T of prepaid credit is set to avoid unnecessary force-termination of the in-progress IMS-to-PSTN calls. The output measures to be optimized are the unnecessary force-termination probability P_{UFT} and the expected extra delay $E[t_d]$ for message delivery. We investigated how these two output measures are affected by input parameters including the expected call holding time $E[t_c]$, the variance V_c of the call holding time, the instant message arrivals rate λ_m, the initial prepaid credit X and the threshold X_T. We made the following observations:

- P_{UFT} increases as λ_m increases. This effect becomes insignificant when X_T is large (e.g., $X_T \geq 6E[t_c]$ in our examples). On the other hand, $E[t_d]$ is always insignificantly affected by λ_m.
- Both P_{UFT} and $E[t_d]$ decrease as X increases.
- When X_T is small, increasing X_T reduces P_{UFT} significantly. When X_T is large (e.g. $X_T > 5E[t_c]$ in our examples), increasing X_T only has insignificant impact on P_{UFT}. On the other hand, increasing X_T always increases $E[t_d]$.
- Both output measures P_{UFT} and $E[t_d]$ degrade as V_c increases.

This appendix provides guidelines to select an appropriate X_T. Based on the above study and observations, a telecom operator should select appropriate X_T values for various traffic conditions.

B.5 Notation

- $1/\lambda$: the expected interval between when the previous prepaid call completes and when the next prepaid call arrives
- λ_m: the arrival rate of the instant messages
- P_{UFT}: the UFT (unnecessary force-termination) probability of an in-progress call in the prepaid application server
- t_c: the prepaid call holding time
- t_d: the extra delay between when an instant message service request arrives and when the instant message is delivered
- T_m: the charge for a prepaid instant message delivery
- V_c : the variance of the call holding time t_c
- X: the amount of the initial prepaid credit
- X_T : the credit threshold that avoids force-termination of prepaid calls
- x^*: the remaining credit in a user's prepaid account

References

[3GP06a] 3GPP, 3rd Generation Partnership Project; Technical Specification Group Core Network; IP Multimedia Subsystem (IMS); Stage 2 (Release 5), 3G TS 23.228 version 5.15.0 (2006-06), 2006.

[3GP06b] 3GPP, 3rd Generation Partnership Project; Technical Specification Group Core Network and Terminals; Technical realization of the Short Message Service (SMS) (Release 6), 3G TS 23.040 V6.8.1 (2006-10), 2006.

[IET96] IETF, RTP: A Transport Protocol for Real-Time Applications. IETF RFC 1889, 1996.

[IET02a] IETF, Session Initiation Protocol (SIP) Extension for Instant Messaging. IETF RFC 3428, 2002.

[IET02b] IETF, SIP: Session Initiation Protocol. IETF RFC 3261, 2002.

[Lin01] Lin, Y.-B. and Chlamtac, I., *Wireless and Mobile Network Architectures*. John Wiley & Sons, Ltd., Chichester, UK, 2001.

[Rao03] Rao, H. C.-H., Cheng, Y.-H., Chang, K.-S. and Lin, Y.-B., iMail: A WAP mail retrieving system, *Information Sciences*, **151**:71–91, 2003.

[Sou05] Sou, S.-I., Wu, Q., Lin, Y.-B. and Yeh, C.-H., Prepaid mechanism of VoIP and messaging services. *IEEE International Conference on Information Technology Research and Education* (*ITRE*), 2005.

[Sou07] Sou, S.-I., Lin, Y.-B., Wu, Q. and Jeng, J.-Y., Modeling prepaid application server of VoIP and messaging services for UMTS, *IEEE Transactions on Vehicular Technology*, **56**(3):1434–1441, 2007.

[Vovida] http://www.vovida.org.

Appendix C

Modeling Credit Reservation for OCS

Through the credit reservation mechanism, the OCS provides real-time charging control in UMTS/IMS network. This appendix investigates the performance of credit reservation for OCS. We assume that a mobile user may access n types of session-based IMS services. After an online service user has purchased some credit units, she is allowed to enjoy multiple sessions simultaneously. Each service type has its own traffic characteristics and communications parameters. For example, the average call holding time for a *Voice over IP* (VoIP) call session is 1–3 minutes (see Chapter 10), and the average session holding time for the interactive mobile gaming sessions may range from 10 to 30 minutes. Details of the UMTS/IMS service characteristics and communications parameters can be found in Section 3.11. Note that the service sessions can be charged according to time duration or packet volume. For a session of type i, each time the OCS grants θ_i credit units to the session. When these credit units are consumed, the OCS grants next θ_i credit units to the session through the Diameter credit reservation procedure. When the balance of the user account at the OCS is below a recharge threshold C_{min}, the OCS does not allow the user to initiate new sessions, and reminds the user to refill the prepaid account by sending a recharge message. This appendix shows how to select an appropriate recharge threshold C_{min}. Note that the usage of C_{min} is different from that of X_T in Appendix B. The reader is encouraged to distinguish these two thresholds and see if they can be integrated. The notation used in this appendix is listed in Section C.3.

C.1 Recharge Threshold-based Credit Reservation

The Diameter credit reservation procedure exercised between an IMS *Application Server* (AS) and the OCS is described in Figure 5.9 of Section 5.5.2. In the *Reserve*

Charging for Mobile All-IP Telecommunications Yi-Bing Lin and Sok-Ian Sou

Units /Debit Units (RU/DU) operations (i.e., Steps 1 or 3 of this procedure), the IMS AS sends the CCR messages to the OCS to request extra credit. In both Steps 2 and 4 of the procedure, the OCS reserves extra credit units for the IMS AS. Specifically, the OCS needs to determine how to allocate the credit units when the remaining credit left in the prepaid account is too small, and when to send the recharge message. These issues can be addressed by a simple mechanism called *Recharge Threshold-based Credit Reservation* (RTCR). In this mechanism, when the amount of the remaining prepaid credit in the OCS is less than a recharge threshold C_{min}, the OCS reminds the user to refill the prepaid account by sending a recharge message to the user's mobile device. Also, the OCS will reject new service requests, and only allow the existing service sessions to consume the remaining credit. We say that the RCTR execution "ends" if all in-progress sessions are complete or force-terminated after the recharge threshold C_{min} is reached.

In RTCR, if C_{min} is set too small, then the amount of the remaining credit may not be large enough to support all in-progress service sessions, and some of them will be forced to terminate. Clearly, force-termination degrades user satisfaction. For example, when all credit units are consumed during an interactive multimedia game, the user is forced to stop playing the game.

On the other hand, if C_{min} is set too large, the OCS will inappropriately reject new session requests that should be accommodated and can be completed before the user's credit is actually depleted. The user will be frequently asked to refill the prepaid account while she still has enough credit. Therefore, C_{min} should be appropriately selected to balance unnecessary force-termination against frequent recharging. Three output measures for RTCR are described below:

- $E[N_i]$: the expected number of the RU/DU operations executed during a type-i session. The larger the $E[N_i]$ value, the higher the credit control message overhead.
- P_c: the completion probability that all in-progress sessions can be finished. The larger the P_c value, the better the user satisfaction.
- $E[C_d]$: the expected amount of unused credit units in the user account when the RTCR execution ends. Note that $C_d = 0$ if any in-progress session is forced to terminate at the end of RTCR execution. It is apparent that the smaller the $E[C_d]$ value, the better the credit utilization in the user account.

Based on the study in [Sou07], the next section uses numerical examples to demonstrate the impact of parameter C_{min} on the RTCR performance.

C.2 Numerical Examples and Conclusions

This section investigates the performance of the RTCR mechanism, including the output measures $E[N_i]$, P_c, and $E[C_d]$. The input parameters are the amount θ_i of credit units that the OCS grants in each RU/DU operation for a type-i session, the recharge threshold C_{min}, and the number n of the service types. We make the following assumptions:

- The inter-session arrivals of type-i service are a Poisson process with the arrival rate λ_i.
- The session holding time of type-i service has a Gamma distribution with the mean $1/\mu_i$ and the variance V_i.

For the discussion purpose, we define the expected session holding time $1/\mu$ (by averaging session holding times of all service types). In our study, both C_{min} and $E[C_d]$ are normalized by $1/\mu$.

Effects of θ_i on $E[N_i]$. Figure C.1 plots the expected number $E[N_i]$ of RU/DU operations executed in a type-i service session against the granted credit θ_i. Note that $E[N_i]$ is not affected by C_{min} and n. This figure shows a trivial result that $E[N_i]$ decreases as θ_i increases. A non-trivial observation is that when $\theta_i \geq 2.5/\mu_i$, $E[N_i] \approx 1$. It implies that selecting θ_i value larger than $2.5/\mu_i$ will not improve the $E[N_i]$ performance.

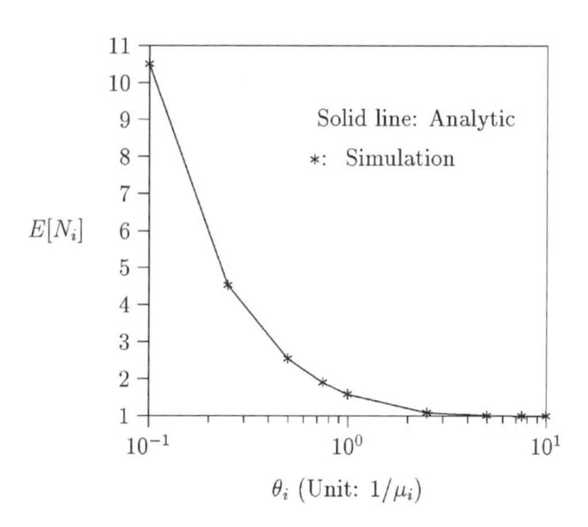

Figure C.1 Effect of θ_i on $E[N_i]$

Effects of C_{min}. Figure C.2 plots the completion probability P_c and the expected credit $E[C_d]$ against θ_i and C_{min}, where $n=2$, $\lambda_1=\mu_1$ and $\lambda_2=\mu_2=2\mu_1$. Figure C.2(a) shows that P_c increases as C_{min} increases. By increasing C_{min}, more unused credit units are available in the prepaid account. Therefore, the possibility of force-termination reduces. For $\theta_i = 1/\mu_i$, when C_{min} increases from $2/\mu$ to $4/\mu$, P_c increases from 81.22% to 95.01%. On the other hand, Figure C.2(b) shows that $E[C_d]$ increases as C_{min} increases. It is apparent that when the threshold C_{min} is reached at the OCS, the exact unused credit units (i.e., C_{min} plus the credits already reserved for the

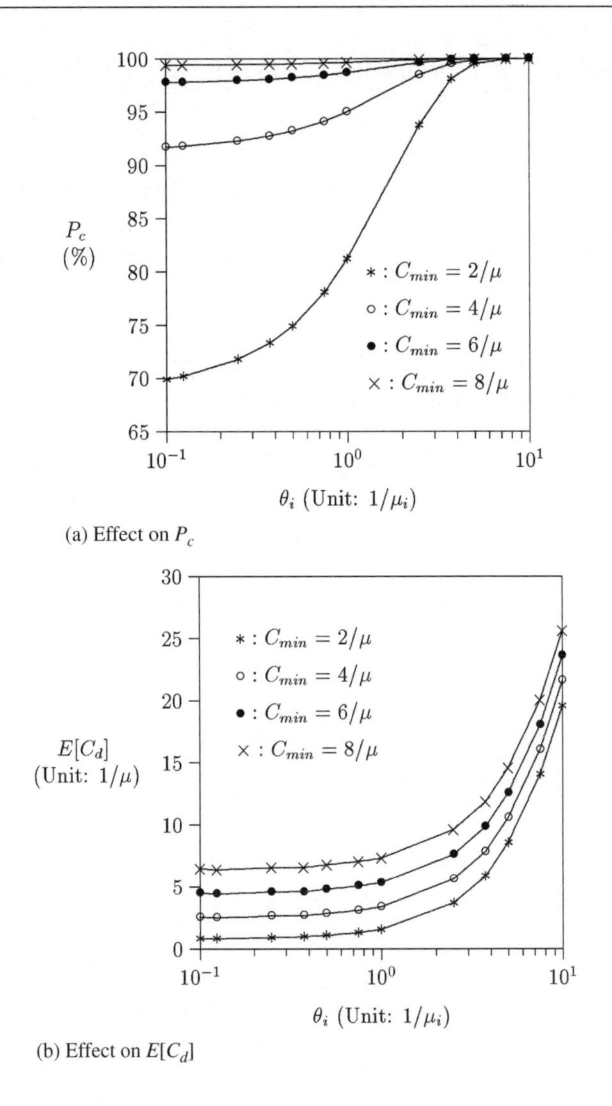

(a) Effect on P_c

(b) Effect on $E[C_d]$

Figure C.2 Effects of θ_i and C_{min} ($n = 2$, $\lambda_1 = \mu_1$ and $\lambda_2 = \mu_2 = 2\,\mu_1$)

in-progress sessions) for the user increases as C_{min} increases. Thus, the amount of consumed credit reduces, and the expected credit $E[C_d]$ increases. For $\theta_i = 1/\mu_i$, when C_{min} increases from $2/\mu$ to $4/\mu$, $E[C_d]$ increases from $1.60/\mu$ to $3.41/\mu$. In this scenario, we expect that $3.41–1.60 = 1.81$ more sessions are complete when $C_{min} = 2/\mu$ than when $C_{min} = 4/\mu$.

Effects of θ_i. Figure C.2(a) shows that P_c is an increasing function of θ_i. When θ_i increases, more credit units are granted to the service session. Therefore, the possibility of force-termination reduces. For $C_{min} = 2/\mu$, when θ_i increases from $1/\mu_i$ to $2.5/\mu_i$, P_c increases from 81.21% to 93.77%. This effect becomes insignificant when

θ_i is large (e.g., $\theta_i \geq 5/\mu_i$). Figure C.2(b) shows that $E[C_d]$ is an increasing function of θ_i. For $C_{min} = 6/\mu$, when θ_i increases from $1/\mu_i$ to $2.5/\mu_i$, $E[C_d]$ increases from $5.37/\mu$ to $7.64/\mu$. Figure C.2(b) also quantitatively indicates how the θ_i and the C_{min} values affect $E[C_d]$. When $\theta_i \leq 1/\mu_i$, $E[C_d] \approx C_{min}$. On the other hand, $E[C_d] >> C_{min}$ as θ_i increases. For example, when $C_{min} = 6/\mu$ and $\theta_i = 10/\mu_i$, $E[C_d] = 23.63/\mu >> 6/\mu$.

Effects of V_i. Figure C.3 plots P_c and $E[C_d]$ against C_{min} and the variance V_i of the Gamma session holding time t_i, where $n = 2$, $\theta_i = 2.5\mu_i$, $\lambda_1 = \mu_1$ and $\lambda_2 = \mu_2 = 2\mu_1$. Figure C.3(a) shows that P_c decreases as V_i increases. This phenomenon is explained

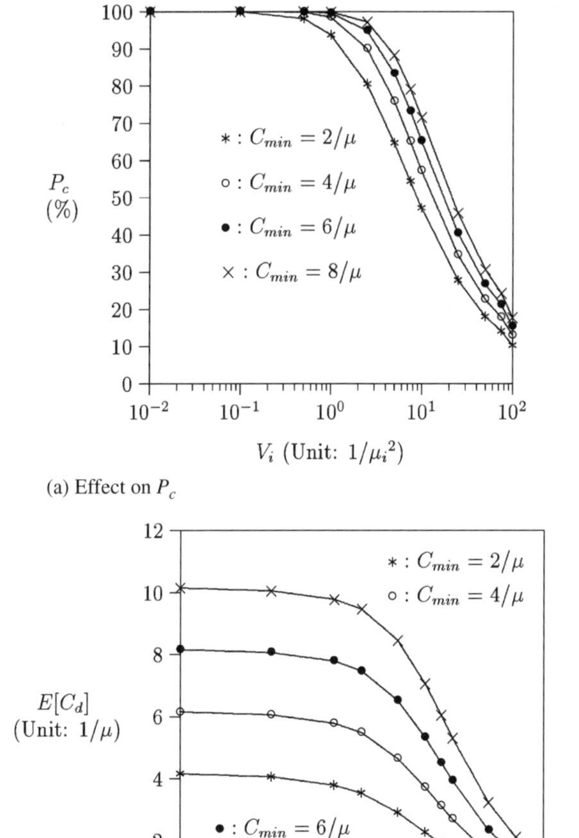

Figure C.3 Effects of V_i ($n = 2$, $\theta_i = 2.5\mu_i$, $\lambda_1 = \mu_1$ and $\lambda_2 = \mu_2 = 2\,\mu_1$)

as follows: As V_i increases, more long and short t_i periods are observed. The recharge messages are more likely to be sent in the long t_i periods than the short t_i periods, and larger residual session holding times $\bar{t}_i$ are expected. Therefore, P_c decreases as V_i increases. Figure C.3(b) also shows that $E[C_d]$ decreases as V_i increases. As V_i increases, the recharge messages are likely to be sent in the long t_i periods. In this case, the possibility that $\bar{t}_i \geq C_{min}$ (i.e., $C_d = 0$) increases. Therefore $E[C_d]$ decreases as V_i increases.

Effects of *n*. Figure C.4 plots P_c and $E[C_d]$ against θ_i and the number n of the service session types, where $C_{min} = 6/\mu$, $\lambda_i = \mu_i = i\mu_1$. Figure C.4(a) shows that P_c

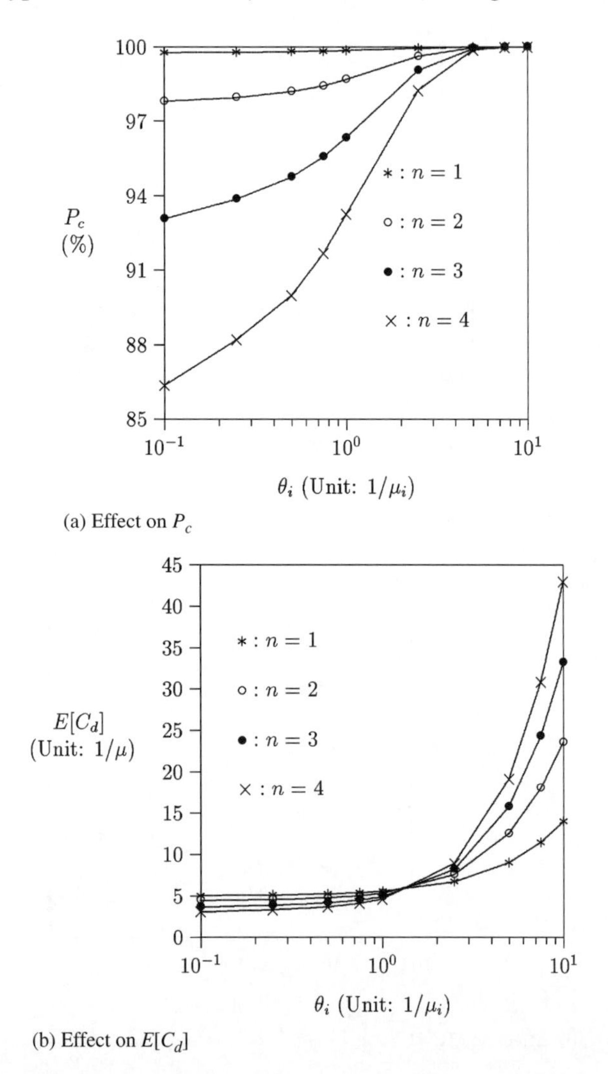

Figure C.4 Effects of n ($C_{min} = 6/\mu$, $\lambda_i = \mu_i = i\,\mu_1$)

is a decreasing function of n. As n increases, the number of in-progress service sessions increases when the recharge message is sent (i.e., when the C_{min} threshold is reached), and therefore P_c decreases. In Figure C.4(b), when θ_i is small (e.g., $\theta_i \leq 1/\mu_i$), $E[C_d]$ decreases as n increases. On the other hand, when θ_i is large (e.g., $\theta_i \geq 2.5/\mu_i$), $E[C_d]$ increases as n increases. This phenomenon is explained as follows: When the recharge message is sent, the number of simultaneous in-progress service sessions increases as n increases. Two conflicting effects are observed when n increases. First, more credit units will be consumed in these sessions and $E[C_d]$ decreases. Second, the "net" unused credit units which have been granted to the service sessions increase, and $E[C_d]$ increases. When θ_i is small, the first effect dominates. On the other hand, the second effect dominates when θ_i is large.

C.3 Notation

- C_{min}: the recharge threshold in RTCR
- $E[C_d]$: the expected amount of unused credit units in the user account at the end of RTCR execution (before recharging)
- $E[N_i]$: the expected number of the RU/DU operations executed during a type-i session
- λ_i: the inter-session arrival rate of type-i service
- $1/\mu_i$: the expected session holding time of type-i service
- $1/\mu$: the expected session holding time (by averaging the session holding times of all service type)
- n: the number of types of session-based IMS services
- P_c: the completion probability that all in-progress sessions can be finished (for all service types)
- θ_i: the amount of credit units that the OCS grants in each RU/DU operation for a type-i session
- t_i: the holding time of the type-i service session
- $\bar{t}_i$: the residual holding time of the type-i service session
- V_i: the variance of the service session holding time t_i

Reference

[Sou07] Sou, S.-I., Hung, H.-N., Lin, Y.-B., Peng, N.-F. and Jeng, J.-Y., Modeling credit reservation procedure for UMTS online charging system, *IEEE Transactions on* Wireless *Communications*, **6**(11): 4129–4135, 2007.

Appendix D

Reducing Credit Re-authorization Cost

This appendix studies how to reduce credit re-authorization cost in the OCS. As described in Chapter 8, the *Session Based Charging Function* (SBCF; see Figure 8.3(a)) in the OCS is responsible for online charging of network bearer and user sessions [3GP06b]. The SBCF interacts with the *Account Balance Management Function* (ABMF; see Figure 8.3(c)) to query and update the user's account through the Rc interface. The ABMF keeps the subscriber's account data and manages the account balance. At the time of writing this book, the message exchanges in the Rc interface are not defined. In this appendix, we use ABMF Request and ABMF Response to represent the messages exchanged between the SBCF and the ABMF. The SBCF interacts with the *Rating Function* (RF; see Figure 8.3(e)) to determine the price and tariff of the requested service through the Re interface. By using Diameter Tariff Request/Response and Price Request/Response messages, the RF handles a wide variety of rateable instances, such as data volume and session connection time. The CDRs generated by the SBCF are then transferred to the *Charging Gateway Function* (CGF; see Figure 8.3(d)) via the Ga interface. The CGF acts as a gateway between the 3GPP network and the billing system. In online GPRS sessions, the Diameter credit control protocol is used for communications between the OCS and the TPF/GGSN described in Chapter 9. The OCS credit control is achieved by exchanging the Diameter Credit Control Request (CCR) and the Credit Control Answer (CCA) messages described in Chapter 5.

In a telecom network, the ABMF and the SBCF may physically reside at different (and possibly very remote) locations. Therefore, the message exchanges in the Rc interface may be expensive, and it is desirable to reduce the credit re-authorization message cost.

Charging for Mobile All-IP Telecommunications Yi-Bing Lin and Sok-Ian Sou

D.1 Credit Re-authorization Procedure

Consider a scenario where a mobile user is viewing a streaming video through the GPRS network [3GP06a]. The streaming services are charged according to the amount of time units and the provisioned QoS (i.e., the bandwidth allocated to the GPRS session). Due to user mobility between different UMTS coverage areas, and depending on the workload of the radio network, the QoS of the streaming session (a GPRS session) may change from time to time. In terms of bandwidth allocated, we assume that there are N QoS classes for a GPRS session. For $1 \leq i \leq N$, let α_i be the number of credit units charged for every time unit in QoS class i session, and the OCS granted $\alpha_i \tau_g$ credit units (which can last for τ_g time units) to the GPRS session in each credit reservation. Note that the bandwidth allocated to a class i session is typically proportioned to the charge per time unit (i.e., α_i). The reader should also note the difference (or similarity) between N and the variable n defined in Appendix C.

Whenever the QoS of the GPRS session changes, the credit re-authorization procedure illustrated in Figure D.1 (based on Figure 8.9) must be executed. Note that in Figure 8.9, the AMBF is a component of the OCS. In Figure D.1, the ABMF is separated from the OCS to emphasize that this component is placed in a remote location.

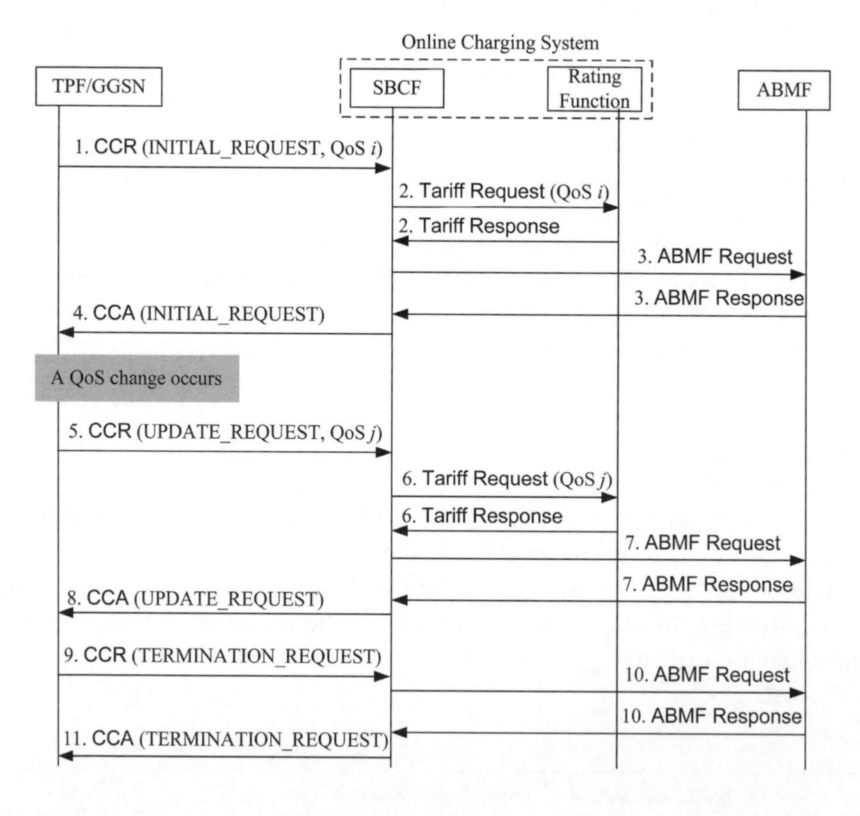

Figure D.1 Message flow for credit re-authorization procedure (the ABMF is placed remotely from other components of the OCS)

Step 1. To start the streaming GPRS session with online charging, the TPF/GGSN sends the INITIAL_REQUEST CCR message to the OCS. This message indicates the QoS parameter (e.g., QoS class i) for the session.

Step 2. When the OCS receives the credit control request, the SBCF sends the Tariff Request message (including the QoS parameter in the *Service-Information* field) to the RF. The RF replies with the Tariff Response message to indicate the applicable tariff α_i for this session.

Step 3. Based on the received tariff information, the SBCF grants $\alpha_i \tau_g$ credit units to the session. By exchanging the ABMF Request and Response message pair with the ABMF, the SBCF reserves $\alpha_i \tau_g$ credit units from the subscriber's account.

Step 4. After the reservation is performed, the OCS acknowledges the TPF/GGSN with the CCA message including the $\alpha_i \tau_g$ granted credit units and the trigger event type (i.e., CHANGE_IN_QOS). This message indicates that the TPF/GGSN should trigger credit re-authorization procedure when the QoS change occurs. The GGSN then starts to deliver the service.

Step 5. When the QoS of the session is changed from class i to class j (because, e.g., the mobile user moves to another base station with different bandwidth capacity), the credit re-authorization procedure should be executed. The TPF/GGSN suspends service delivery and sends an UPDATE_REQUEST CCR message to the OCS. This CCR message includes the *Reporting-Reason* field with value RATING_CONDITION_CHANGE and the *Trigger-Type* with value CHANGE_IN_QOS. The TPF/GGSN also reports that $\alpha_i(\tau_g - \tau_u)$ credit units have been consumed, and then requests extra credit units from the OCS based on the QoS parameter (i.e., QoS class j).

Step 6. Upon receipt of the CCR message from the TPF/GGSN, the SBCF calculates the remaining credit units for the GPRS session and revaluates the rating by exchanging Tariff Request and Response messages with the RF. The Tariff Response message includes the new tariff α_j for the session.

Step 7. Based on the old tariff α_i and new tariff α_j for the GPRS session, the SBCF then debits $\alpha_i(\tau_g - \tau_u)$ credit units and requests extra $\alpha_j \tau_g$ credit units from the ABMF.

Step 8. The OCS acknowledges the TPF/GGSN with the CCA message to indicate that $\alpha_j \tau_g$ credit units have been reserved (which lasts for τ_g time units). The GGSN resumes the service delivery. Note that Steps 5–8 are executed whenever the QoS is changed or when the allocated credit units are depleted.

Step 9. When the video streaming service is complete, the GGSN terminates the session. The TPF/GGSN reports the amount of used credit to the OCS through the TERMINATION_REQUEST CCR message.

Step 10. The SBCF calculates the consumed credit units and instructs the ABMF to debit the user account.

Step 11. Finally, the OCS acknowledges the TPF/GGSN with a CCA message.

For the discussion purpose, the above re-authorization procedure (i.e., Steps 6 and 7) in the OCS is referred to as the "basic" scheme. If the QoS of the session changes frequently, the signaling traffic incurred by the re-authorization procedure becomes significant. In [Sou08], we proposed a "threshold-based" scheme that utilizes a threshold δ to reduce the signaling cost between the SBCF and the ABMF. This new scheme is described in the next section.

D.2 The Threshold-based Scheme

In the basic scheme, the ABMF message exchanges (Step 7 in Figure D.1) can be omitted if the remaining credit units are large enough to accommodate the new reservation for the new QoS. Based on this observation, the threshold-based scheme was proposed to reduce the signaling cost. In this scheme, the SBCF determines whether to interact with the ABMF or not based on a threshold δ. Suppose that a class i GPRS session is allocated $\alpha_i \tau_g$ credit units at time t. At time $t + \tau_g - \tau_u$, the QoS class for this session is changed from class i to class j. In the threshold-based scheme, the ABMF message exchange is skipped if the amount $\alpha_i \tau_u$ of remaining credit is larger than $\delta \alpha_j \tau_g$. Details of this scheme are described as follows:

Step T.1. When Step 6 of Figure D.1 is executed, the SBCF retrieves the old tariff α_i and new tariff α_j for the GPRS session from the RF.

Step T.2a. If $\alpha_i \tau_u \geq \delta \alpha_j \tau_g$, the SBCF directly allocates $\alpha_i \tau_u$ credit units to the GPRS session. The ABMF message exchange (i.e., Step 7 in Figure D.1) is skipped, and the procedure proceeds to Step 8 in Figure D.1.

Step T.2b. Otherwise (i.e., $\alpha_i \tau_u < \delta \alpha_j \tau_g$), the SBCF allocates $\alpha_j \tau_g$ credit units to the GPRS session by executing Step 7 in Figure D.1. That is, the SBCF debits $\alpha_i(\tau_g - \tau_u)$ credit units to the ABMF and makes new reservation for $\alpha_j \tau_g$ credit units. The procedure proceeds to Step 8 in Figure D.1.

In the OCS, a mobile operator (or a user) can check the account balance at any time. When there is no in-progress session, the OCS accurately reports the account balance of the user. On the other hand, when credit units are reserved for some in-progress sessions, the account balance reported by the OCS may not be up to date because some reserved credit units have already been used. When a "balance check" occurs, the OCS reports the account balance including the reserved credit units. This reported value may be larger than the actual balance. Denote C as the inaccuracy of credit information, which is the difference between the balance stored in the OCS (including the reserved credit units) and the actual balance (excluding the credit units already consumed by the GPRS session). In other words, C is the amount of credit consumed by the GPRS session between when the previous ABMF message exchange occurs and when the balance check occurs. In the threshold-based scheme, the inaccuracy of credit information C may be larger than the basic scheme because it skips some

ABMF message exchanges. Therefore, it is important to select appropriate τ_g and δ values to "optimize" the performance of the threshold-based scheme in terms of the following output measures:

- M: the expected number of ABMF message exchanges for a GPRS session. The larger the M value, the higher the ABMF message overhead.
- C: the expected inaccuracy of credit information when a balance check occurs during an in-progress session. The smaller the C value, the more accurate the account balance reported by the OCS.

In our study, the above output measures are subscripted with "B" and "T"; i.e., $M_B(C_B)$, and $M_T(C_T)$, to represent the output measures of the basic scheme and the threshold-based scheme, respectively. We make the following assumptions:

- There are N QoS classes that may change in a GPRS session. A GPRS session starts with QoS class i ($1 \le i \le N$) with probability $1/N$.
- When a QoS change occurs, an in-progress session either terminates (with probability P_0) or switches to another QoS class with probability $(1-P_0)/(N-1)$.
- For $1 \le i \le N$, let α_i be the amount of credit units charged for each time unit in a QoS class i session. For each credit reservation through the ABMF message exchange, the SBCF reserves $\alpha_i \tau_g$ credit units in the ABMF, and then grants these credit units (which lasts for τ_g time units) to the GPRS session, where τ_g is a fixed value.
- The changes of QoS partition a GPRS session into several subsessions. The holding time of each subsession is independent and identically distributed (i.i.d.) exponential random variable with mean $1/\mu_s$.

Following the above assumptions, the reader is encouraged to write a simulation program to compute the output measures M and C.

D.3 Numerical Examples

Based on the study in [Sou08], this section investigates the performance of the OCS credit re-authorization procedure. The following effects are discussed.

Effects of the granted time units τ_g. Figure D.2 plots M_T/M_B and C_T/C_B against the threshold δ and the granted time units τ_g, where $N=2$, $\alpha_2=2\alpha_1$ and $P_0=0.01$. Figure D.2 shows a trivial result that M_T/M_B is a decreasing function of τ_g, and C_T/C_B is an increasing function of τ_g. A non-trivial result is that when $\tau_g \le 5/\mu_s$, the threshold-based scheme significantly reduces the ABMF signaling overhead while the inaccuracy of the credit information insignificantly increases.

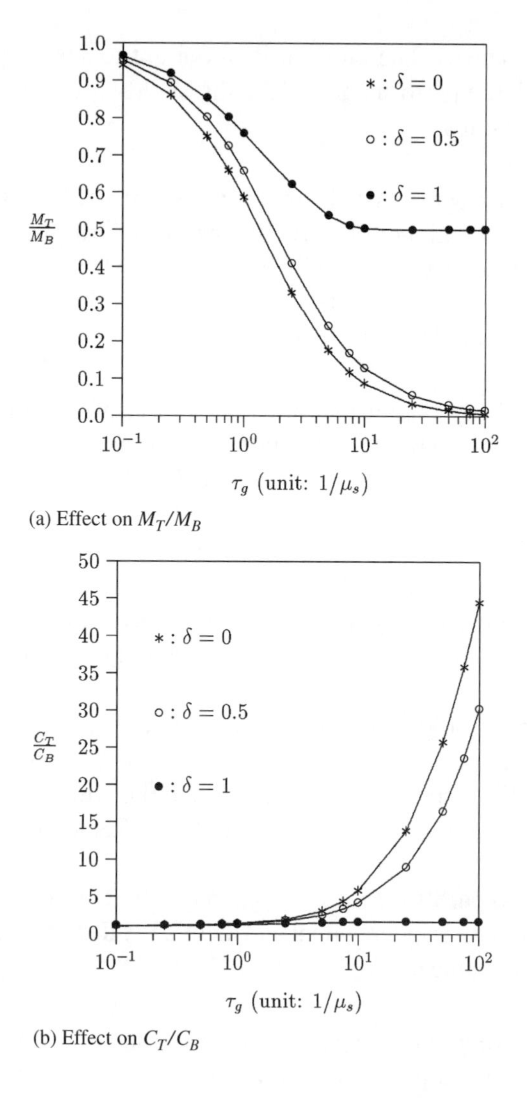

Figure D.2 Effect of $\tau_g (N = 2,\ \alpha_2 = 2\alpha_1$ and $P_0 = 0.01)$

Effects of the threshold δ. Figure D.3 plots M and C against the threshold δ, where $N = 2$, $\alpha_2 = 2\alpha_1$ and $P_0 = 0.01$. The output measures for the basic scheme are not affected by δ. For the threshold-based scheme, M_T increases and C_T decreases as δ increases. When δ is large, the basic scheme and the threshold-based scheme have the same performance. In Figure D.3, the performance of the threshold-based scheme is similar to that of the basic scheme when $\delta \geq 2.5$. Figure D.3 quantitatively indicates how δ and τ_g affect C. In the OCS, the guideline for selecting the granted time units τ_g can be found in Appendix C (note that if $\alpha_i = \alpha_j$ for all i and j, then $\theta_i = \alpha_i \tau_g$ in Appendix C). When a specific τ_g is selected, for example, $\tau_g = 5/\mu_s$, we observe

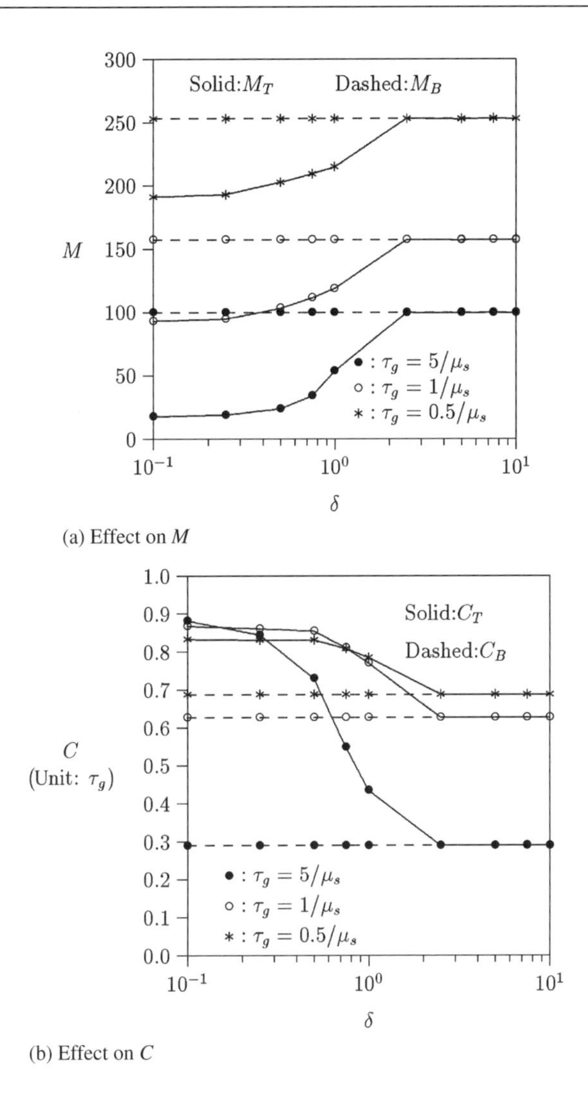

Figure D.3 Effect of $\delta(N = 2,\ \alpha_2 = 2\alpha_1$ and $P_0 = 0.01)$

that $C_B = 0.29\tau_g$ in the basic scheme. If the mobile operator can tolerate a larger inaccuracy of the credit information, e.g., if it is acceptable to select a C value no larger than $5\tau_g$, then we can choose $\delta = 1$ in the threshold-based scheme. In this case, $C_T = 0.43\tau_g$ and the message overhead M is decreased by 46.45% as compared with the basic scheme.

Effects of the session termination probability P_0. Figure D.4 plots the ratios M_T/M_B and C_T/C_B as functions of P_0 and δ, where $N = 2, \alpha_2 = 2\alpha_1$ and $\tau_g = 1/\mu_s$. As P_0 increases, both M_T and M_B decrease. However, the proportion M_T/M_B of the ABMF

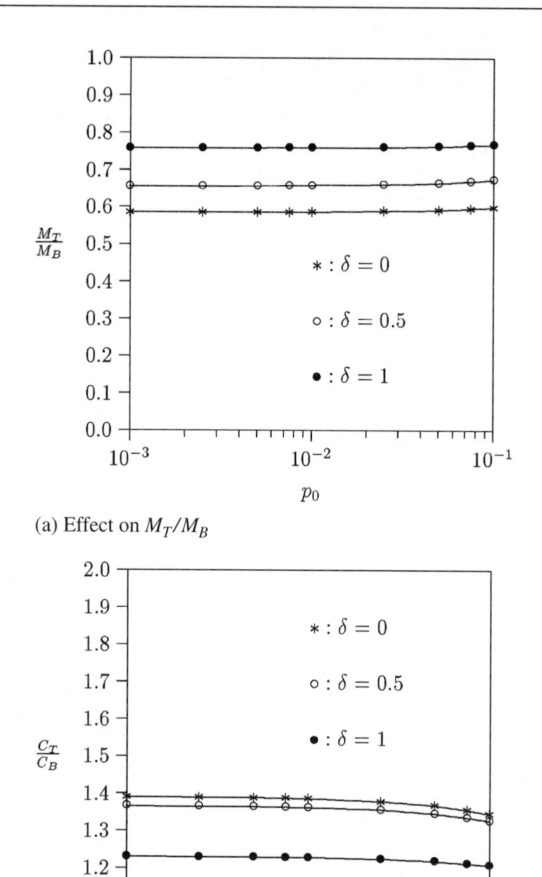

(a) Effect on M_T/M_B

(b) Effect on C_T/C_B

Figure D.4 Effect of δ and $P_0(N = 2, \alpha_2 = 2\alpha_1$ and $\tau_g = 1/\mu_s)$

message exchanges saved by the threshold-based scheme is the same as illustrated in Figure D.4. This figure also shows that C_T/C_B is insignificantly affected by P_0.

Effects of the number N of QoS classes. Figure D.5 plots M_T/M_B and C_T/C_B as functions of N and δ, where $\alpha_i = i\alpha_1$, $P_0 = 0.01$ and $\tau_g = 1/\mu_s$. This figure shows that both ratios M_T/M_B and C_T/C_B are not significantly affected by the number N of QoS classes. For $N = 4$, when δ decreases from 0.5 to 0, M_T/M_B decreases by 9.79% and C_T/C_B only increases by 2.71%. Therefore, it is better to set $\delta = 0$ than $\delta = 0.5$ in this case (note that the threshold-based scheme with $\delta = 0$ means that Step 7 in Figure D.1 is never executed).

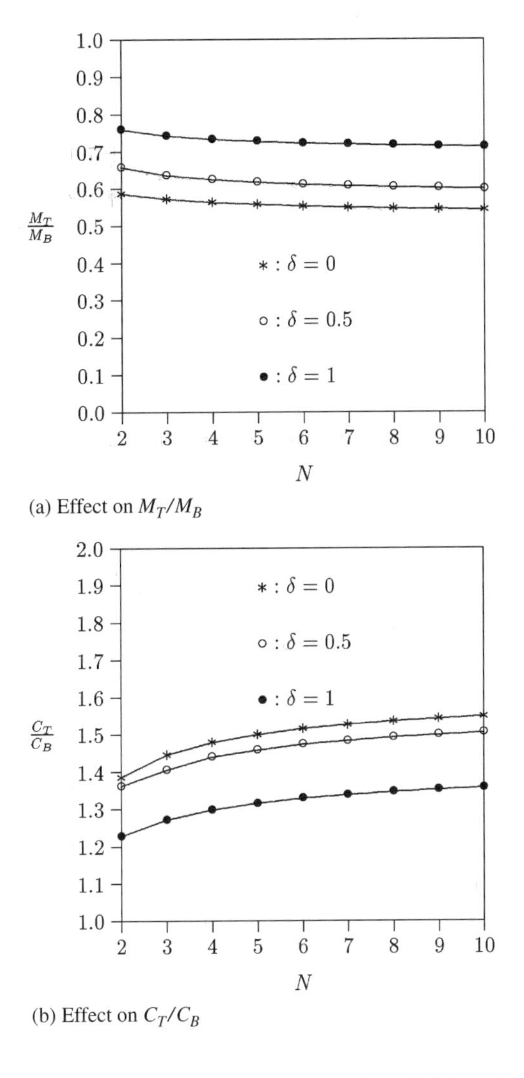

(a) Effect on M_T/M_B

(b) Effect on C_T/C_B

Figure D.5 Effect of δ and N ($\alpha_i = i\alpha_1$, $P_0 = 0.01$ and $\tau_g = 1/\mu_s$)

D.4 Concluding Remarks

This appendix studied credit re-authorization in OCS. We proposed a threshold-based scheme with parameter δ to reduce the signaling traffic for the OCS credit re-authorization procedure. Basically, the threshold-based scheme reduces the number M of ABMF message exchanges during a session at the cost of increasing the inaccuracy of credit information C when a balance check occurs. These two conflicting output measures are affected by the threshold δ and the granted time units τ_g.

We make the following observations (where the subscripts "B" and "T" in the output measures M and C represent the basic scheme and the threshold-based scheme, respectively):

- The ratio M_T/M_B is a decreasing function of τ_g, and the ratio C_T/C_B is an increasing function of τ_g. When τ_g is small, the threshold-based scheme significantly reduces the ABMF signaling overhead while the inaccuracy of the credit information insignificantly increases.
- As the threshold δ increases, M_T increases and C_T decreases. When δ is large, the basic scheme and the threshold-based scheme have the same performance.
- As the session termination probability P_0 increases, both M_T and M_B decrease. However, the ratio M_T/M_B is not affected. Also, the ratio C_T/C_B is insignificantly affected by P_0.
- The ratios M_T/M_B and C_T/C_B are not significantly affected by the number N of QoS classes.

From the above discussion, a telecom operator can select the appropriate δ and τ_g values for various traffic conditions based on our model.

D.5 Notation

- α_i: the number of credit units charged for every time unit in a class i session
- C: the inaccuracy of credit information, which is the difference between the balance stored in the OCS (including the reserved credit units) and the actual balance (excluding the credit units already consumed by the GPRS session)
- $C_B(C_T)$: the C value for the basic scheme (threshold-based scheme)
- δ: the threshold used by the SBCF to determine whether to interact with the ABMF or not
- $1/\mu_s$: the expected holding time of a sub-session
- M: the expected number of ABMF message exchanges for a GPRS session
- M_B (M_T): the M value for the basic scheme (threshold-based scheme)
- N: the number of the QoS classes that may change in a GPRS session
- P_0: the probability that the GPRS session terminates at the end of a sub-session
- τ_g: the time units granted to the GPRS session in each credit reservation
- τ_u: the time units left at the end of the previous sub-session

References

[3GP06a] 3GPP, 3rd Generation Partnership Project; Technical Specification Group Service and System Aspects; Transparent end-to-end Packet-switched Streaming Service (PSS); Protocols and codecs (Release 6), 3G TS 26.234 version 6.9.0 (2006-09), 2006.

[3GP06b] 3GPP, 3rd Generation Partnership Project; Technical Specification Group Service and System Aspects; Telecommunication management; Charging management; Online Charging System (OCS): Applications and interfaces (Release 6), 3G TS 32.296 version 6.3.0 (2006-09), 2006.

[Sou08] Sou, S.-I., Lin, Y.-B. and Jeng, J.-Y., Reducing credit re-authorization cost in UMTS online charging system, *IEEE Transactions on Wireless Communications*, 2008.

Appendix E

Credit Redistribution for UMTS Prepaid Service through CAMEL

In a typical 2G mobile network, a user is allowed to make one prepaid call at a time [Lin01], and the prepaid charging issue is relatively simple. In the 2.5G or the 3G networks, a user may be engaged in multiple voice and/or data prepaid sessions at the same time. To support such services, it is important to distribute an appropriate amount of prepaid credit to the voice calls (in the CS domain) and the data sessions (in the PS domain) that are simultaneously connected to a user. As we described in Appendix C, the network may assign too many prepaid credit units to the already executed prepaid sessions of the user, and therefore no sufficient prepaid credit units are left for new prepaid sessions requested by the same user. On the other hand, the network may assign insufficient amounts of credit to the in-progress prepaid sessions, which results in heavy network traffic for prepaid-related signaling.

This appendix shows how to implement UMTS prepaid services through the *Intelligent Network* (IN) approach, and studies how the prepaid credit allocation mechanism can accommodate more simultaneous prepaid sessions for a user without incurring heavy network signaling. Also, a *Prepaid Credit Reclaim* (PCR) procedure is described to redistribute the credit for new prepaid sessions. With PCR, the network can flexibly allocate the prepaid credit units to simultaneously connected sessions, and thus more prepaid sessions can be accommodated for a user.

E.1 The IN Approach for the UMTS Prepaid Service

Figure E.1 illustrates a simplified UMTS network architecture. In this architecture, the UMTS prepaid service can be supported by CAMEL (see Section 5.1 and 3GPP TS 22.078 [3GP05]) with the following three IN functional entities: gsmSSF, gprsSSF,

Charging for Mobile All-IP Telecommunications Yi-Bing Lin and Sok-Ian Sou

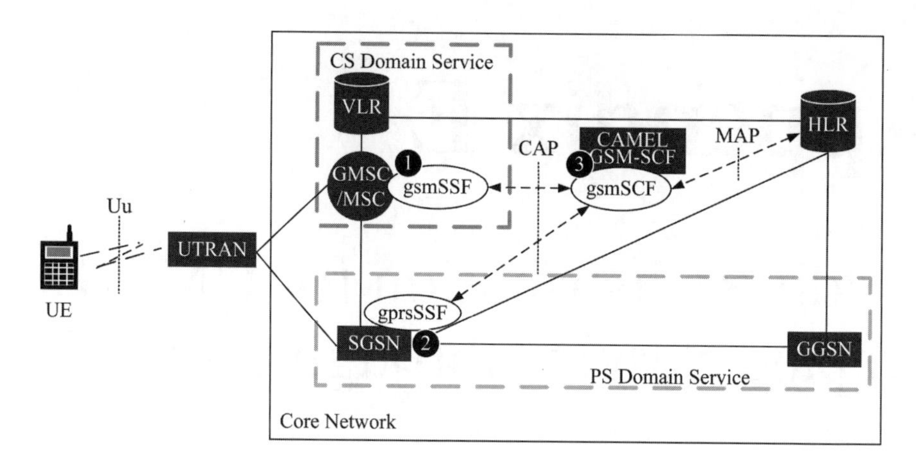

CAMEL: Customized Application for Mobile Network Enhanced Logic
CAP: CAMEL Application Part
HLR: Home Location Register
MSC: Mobile Switching Center
SSF: Service Switching Function
UE: User Equipment
VLR: Visitor Location Register
GGSN: Gateway GPRS Support Node
MAP: Mobile Application Part
SCP: Service Control Point
SGSN: Serving GPRS Support Node
UTRAN: UMTS Terrestrial Radio Access Network

Figure E.1 The UMTS network architecture for prepaid service

and gsmSCF deployed in the MSC (Figure E.1(1)), the SGSN (Figure E.1(2)), and the CAMEL GSM-SCF (Figure E.1(3)), respectively. The gsmSCF instructs the SGSN/MSC to initiate, terminate, suspend or resume an ongoing PS/CS call. The gsmSSF and the gprsSSF communicate with the CAMEL GSM-SCF using the CAP. To set up a CS domain or a PS domain call, a CAP dialogue is established between the CAMEL GSM-SCF and the MSC/SGSN. The CAP messages are exchanged in this dialogue. The MSC/SGSN is responsible for monitoring the call-related events for a CS/PS call (e.g., CS call setup, call termination, PDP context activation, and PDP context deactivation), and forwarding the detected event to the CAMEL GSM-SCF. According to the event type, the CAMEL GSM-SCF instructs the MSC/SGSN to activate IN applications on a connected CS/PS call. For demonstration purposes, we focus on the PS domain prepaid services in this appendix.

When a mobile user subscribes to the prepaid service, an amount of prepaid credit is purchased and is maintained in the CAMEL GSM-SCF. When the mobile user originates a prepaid session, the SGSN reports this event to the CAMEL GSM-SCF. The CAMEL GSM-SCF checks if the user has enough credit to support this session. If so, the CAMEL GSM-SCF assigns some prepaid credit units to the SGSN. These credit units are decremented at the SGSN in real time based on either the traffic volume or the duration time. After the assigned credit units are consumed, the SGSN may ask for more credit units from the CAMEL GSM-SCF. If the credit at the GSM-SCF is depleted, the prepaid session is forced to terminate. To continue the service, the user needs to refill her prepaid credit by purchasing a top-up card.

3GPP specification [3GP05] does not specify the amount of the prepaid credit the CAMEL GSM-SCF should assign to each prepaid session. Since a user may be engaged in several prepaid sessions at the same time as described in Appendix C, it is important to distribute appropriate amounts of prepaid credit to simultaneously connected sessions for a user. Without an intelligent prepaid credit allocation mechanism, the CAMEL GSM-SCF may give too many credit units to the already executed prepaid sessions of the user. If so, insufficient prepaid credit will be left for any newly incoming requests of this user. To resolve this issue, Lin *et al.* [Lin06] proposed a *Prepaid Credit Distribution* (PCD) algorithm for UMTS prepaid services. PCD dynamically adjusts the credit units allocated to simultaneously in-progress prepaid sessions of a user. Furthermore, Lin proposed a *Prepaid Credit Reclaim* (PCR) procedure that can reclaim the credit units of already executed prepaid sessions, and assign the reclaimed credit to newly requested prepaid sessions of the same user. With PCR, the network can flexibly redistribute the prepaid credits to the simultaneously connected prepaid sessions. Thus, more prepaid sessions can be accommodated for a user.

E.2 The Prepaid Charging Message Flow

The UMTS prepaid charging mechanism is specified in 3GPP TS 23.078 [3GP07]. Figure E.2 illustrates the prepaid charging message flow for the volume-based PS domain prepaid services. The CAP prepaid service-related messages are exchanged between the SGSN and the CAMEL GSM-SCF as follows:

Step 1. When the user requests to initiate a new prepaid data session S, the SGSN informs the CAMEL GSM-SCF of this event through the EventReportGPRS message.

Step 2. Upon receipt of the EventReportGPRS message, the CAMEL GSM-SCF checks the amount C_r of the remaining credit units of the user and determines if the session request is granted. If C_r is not large enough to support this request, then Steps 7 and 8 in Figure E.2 are executed to terminate the session (to be elaborated later). Otherwise, the CAMEL GSM-SCF assigns k_1 credit units to session S by sending a message ApplyChargingGPRS(maxTransferredVolume $= k_1$) to the SGSN, where the parameter maxTransferredVolume specifies the amount of credit assigned to this prepaid session. Then the CAMEL GSM-SCF decrements the remaining credit by k_1 units; i.e., $C_r \leftarrow C_r - k_1$. We assume that the amount k_1 is not a fixed value, and is determined by the PCD algorithm (to be described later). When k_1 is a fixed value, its value is equivalent to θ in Appendix C.

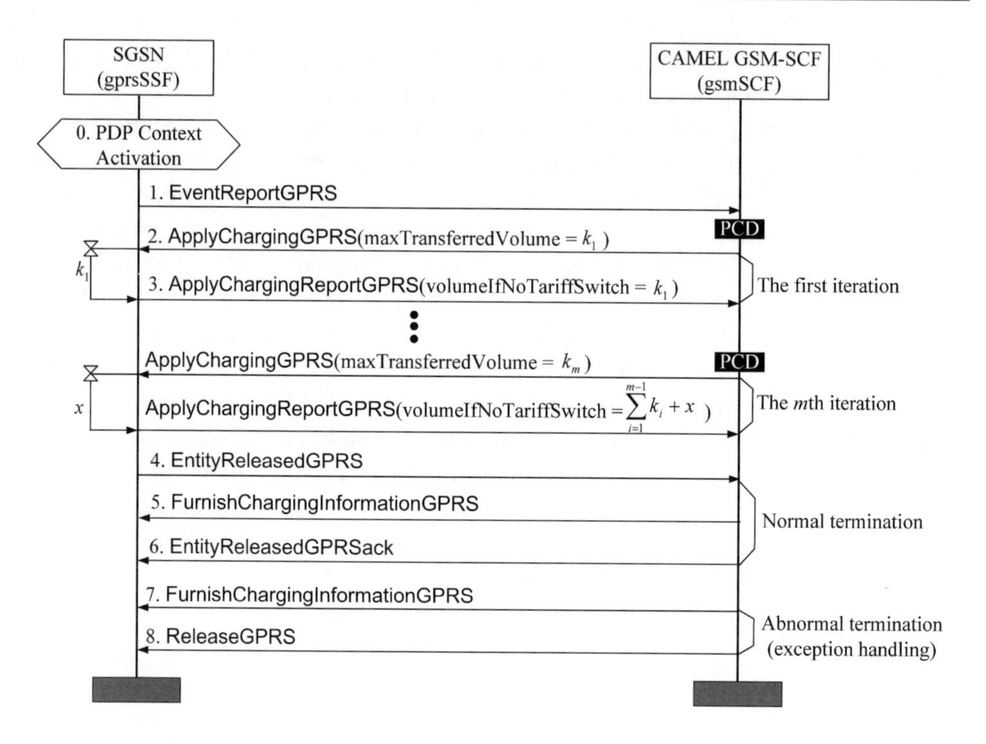

Figure E.2 Message flow for the volume-based prepaid charging

Step 3. After k_1 credit units have been exhausted, the SGSN suspends packet transmission for S, and sends the ApplyChargingReportGPRS (volumeIfNoTariffSwitch = k_1) message to the CAMEL GSM-SCF to indicate that k_1 credit units have been consumed, and more credit units are required to serve this session. In this message, the parameter volumeIfNoTariffSwitch specifies the amount of credit units that have been consumed by S so far.

Define an *iteration* as a consecutive execution of Steps 2 and 3, which is equivalent to an RU operation described in Appendix C. Assume that S is normally terminated after m iterations, and that at the mth iteration, the SGSN consumes x credit units where $x \leq k_m$. The volumeIfNoTariffSwitch parameter is set to $\sum_{i=1}^{m-1} k_i + x$ in the mth ApplyChargingReportGPRS message. Upon receipt of the mth ApplyChargingReportGPRS message, the CAMEL GSM-SCF updates the amount of remaining credit as $C_r \leftarrow C_r + (k_m - x)$.

The SGSN completes the session by executing the following steps.

Step 4. By sending the EntityReleasedGPRS message, the SGSN informs the CAMEL GSM-SCF that session S has been terminated.

Step 5. Upon receipt of the EntityReleasedGPRS message, the CAMEL GSM-SCF replies with a FurnishChargingInformationGPRS message to the SGSN. This message includes the CDR for this session.

Step 6. The CAMEL GSM-SCF sends the EntityReleasedGPRSack message to the SGSN, and charging for the session is terminated.

Suppose that after the ith iteration, C_r in the CAMEL GSM-SCF is not large enough to support the i+1st iteration (i.e.,$C_r < k_{i+1}$). Then the following two steps are executed.

Step 7. This step is the same as Step 5. The CAMEL GSM-SCF sends a FurnishChargingInformationGPRS message that provides the CDR to the SGSN.

Step 8. The CAMEL GSM-SCF sends a ReleaseGPRS message to the SGSN to terminate the ongoing session.

At Step 2, the amount k_i may significantly affect the number of simultaneously executed prepaid sessions and the number of the messages exchanged during a prepaid session. If k_i is large, it is more likely to complete a session before k_i credit units are consumed, and a small number of prepaid signaling messages are exchanged. However, if the CAMEL GSM-SCF gives too many credit units to the already executed prepaid sessions of the user, there may not be enough remaining credit units to support new session requests. Therefore, the amounts of credit should be carefully allocated. To flexibly assign the credit units to the prepaid sessions, Lin *et al.* [Lin06] proposed the *Prepaid Credit Distribution* (PCD) algorithm. This algorithm is executed at the CAMEL GSM-SCF before the ApplyChargingGPRS message is sent (see the black boxes in Figure E.2). The PCD algorithm is illustrated in Figure E.3.

At the beginning of Step 2 for the ith iteration, the PCD attempts to assign θ credit units to the requested prepaid session (by setting j to 0). If $C_r > \theta$, then $k_i = \theta$, and

```
Algorithm PCD(θ, n,γ);

1    begin
2        j ← 0;
3        while j ≤ n do
4            if C_r ≥ θγ^j then
5                return k_i ← θγ^j;
6            else j ← j+1;
7        return k_i ← 0;
8    end
```

Figure E.3 Prepaid Credit Distribution (PCD)

the assignment is successful. Otherwise, a reduction factor $\gamma < 1$ is used to reduce the k_i value. Specifically, PCD attempts to find the largest j such that $C_r \geq \theta\gamma^j$ (Step 5 in Figure E.3). Let n be the maximum number of reductions for the k_i value. If $j > n$, then the k_i assignment fails (Step 7 in Figure E.3). Otherwise, k_i is set to $\theta\gamma^j$.

Note that if the prepaid sessions for a mobile user are controlled by the same SGSN, the prepaid credit may be adjusted locally at the SGSN. However, the interactions between the SGSN and the CAMEL GSM-SCF may need to be modified. Furthermore, the simultaneously requested sessions are likely to be issued from different domains (e.g., SGSN, MSC, WLAN gateway, VoIP call server, etc.), and the prepaid credit units are decremented by the gateways in these domains in a distributed manner. In this case, locally adjusting prepaid credit within a node does not work. Also, a prepaid account can be shared by several members. In this case, multiple SGSNs (for several authorized members) may simultaneously consume the same prepaid credit source.

E.3 The Prepaid Credit Reclaim (PCR) Mechanism

This section describes the *Prepaid Credit Reclaim* (PCR) mechanism. When the execution of PCD fails (i.e., the remaining credit is not large enough to support a new session request or a served prepaid session), the PCR mechanism is triggered to reclaim credit units from the already served sessions. We note that in this scenario, all served sessions (and the new session) are handled by different SGSNs (WLAN gateways or MSCs).

Figure E.4 illustrates the message flow of PCR which consists of four steps.

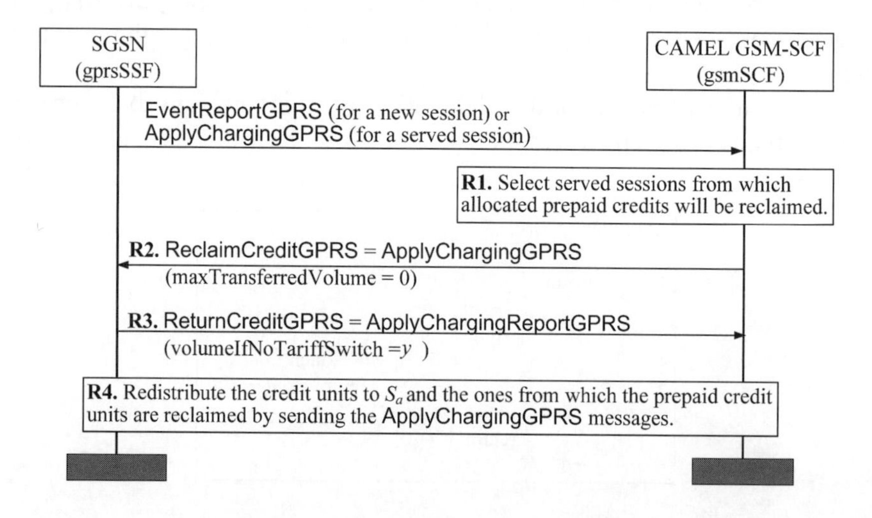

Figure E.4 Message flow for PCR

Suppose that the user makes a new prepaid session request S_a (i.e., an EventReportGPRS message is received by the CAMEL GSM-SCF) or the SGSN sends an ApplyChargingReportGPRS message to request more credit to continue packet transmission for a served prepaid session S_a. Assume that C_r at the GSM-SCF is not large enough to support the request, i.e., the execution of PCD fails. Then PCR is executed as follows:

Step R1. This step selects some currently served prepaid sessions, from which the CAMEL GSM-SCF may reclaim credit units to serve the session S_a. Different selection policies, e.g., random or the-largest-first can be adopted. If no served session is found, the PCR mechanism exits, and S_a is rejected. Let $S_1, S_2, S_3, \ldots, S_L$ be prepaid sessions (handled by different SGSNs/MSCs/WLAN gateways; without loss of generality, the remainder of this appendix uses the term SGSN to represent the network nodes) selected for credit reclaim, where L is the number of the selected sessions.

Step R2. Suppose that Session S_l is handled at SGSN $l (1 \leq l \leq L)$. For every l, the CAMEL GSM-SCF sends a ReclaimCreditGPRS message to SGSN l. This message can be implemented by the CAP message ApplyChargingGPRS (maxTransferredVolume = 0), which informs SGSN l that the assigned credit units of S_l will be reclaimed.

Step R3. Upon receipt of the ReclaimCreditGPRS message, SGSN l suspends packet transmissions for S_l, and replies to the CAMEL GSM-SCF with a ReturnCreditGPRS message (which can be implemented by the CAP message ApplyChargingReportGPRS) to indicate the amount y of returned credit through parameter volumeIfNoTariffSwitch.

Step R4. After the CAMEL GSM-SCF receives all ApplyChargingReportGPRS messages, PCR redistributes the reclaimed credit units to $S_a, S_1, S_2, \ldots, S_L$. Suppose that the amount of the remaining credit is C_r. If $C_r/(L+1)$ is less than a threshold $\varepsilon > 0$, then the reclaimed credit units are returned to $S_1, S_2, \ldots, S_L$ by sending the ApplyChargingGPRS message. Session S_a is rejected, and PCR exits. Otherwise (i.e., $C_r/(L+1) \geq \varepsilon$), the CAMEL GSM-SCF sends the message ApplyChargingGPRS (maxTransferredVolume = $C_r/(L+1)$) to SGSN l to support the session S_l.

Two reclaim-related CAMEL message types are introduced in PCR: ReclaimCredit GPRS and ReturnCreditGPRS. These two message types can be implemented by reusing two existing CAP message types: ApplyChargingGPRS and ApplyChargingReportGPRS, as described in Steps R1 and R2. To trigger PCR, we add reclaim logic functions in the CAMEL GSM-SCF and the SGSN. When the CAMEL GSM-SCF receives an EventReport message and finds that the user does not have enough credit to support this new prepaid session, the reclaim logic in the CAMEL

GSM-SCF is triggered to send the ReclaimCreditGPRS message. When the SGSN receives the ReclaimCreditGPRS message, the reclaim logic of the SGSN suspends the current data transmission, and returns the remaining prepaid credit through the ReturnCreditGPRS message. It is clear that the implementation of the PCR procedure does not modify the CAP protocol and only requires minor modifications to the SGSN and the CAMEL GSM-SCF. If the reclaim logic is not implemented in the SGSN, then the ReclaimCreditGPRS message is ignored when the SGSN receives this message. In other words, the PCR approach is backward compatible where credit reclaiming is only exercised when the SGSN is equipped with PCR, and the standard 3GPP TS23.078 procedure is exercised if PCR is not installed in the SGSN.

E.4 Concluding Remarks

This appendix investigated the prepaid services for the 2.5G or the 3G networks where multiple voice and/or data prepaid sessions are simultaneously supported for a user. We described the UMTS prepaid network architecture based on the Intelligent Network (IN) protocol, and a Prepaid Credit Distribution (PCD) algorithm that dynamically allocates the amount of credit to support simultaneously prepaid sessions of a user. Then we elaborated on a Prepaid Credit Reclaim (PCR) mechanism that reclaims credit units from the currently served prepaid sessions. This mechanism provides extra flexibility for credit allocation.

In [Lin06], simulation experiments were conducted to investigate the performance of the PCD and the PCR mechanisms. The study indicated that with PCR, more prepaid sessions can be simultaneously accommodated for a user as compared with PCD. As a final remark, PCR is a pending patent issued from Chunghwa Telecom, Taiwan.

E.5 Notation

- C_r: the amount of the remaining credit units of a user
- k_i: the amount of the credit assigned to a GPRS session in the ith iteration
- θ: the maximum amount of credit that the PCD attempts to assign to session S in an iteration
- S, S_a, S_l: a prepaid GPRS data session

References

[3GP05] 3GPP, 3rd Generation Partnership Project; Technical Specification Group Services and System Aspects; Customized Applications for Mobile network Enhanced Logic (CAMEL); Service description; Stage 1 (Release 5), 3G TS 22.078 version 5.15.0 (2005-03), 2005.

[3GP07] 3GPP, 3rd Generation Partnership Project; Technical Specification Group Core Network; Customized Applications for Mobile network Enhanced Logic (CAMEL); Phase X, Stage 2. Technical Report 3G TS version 23.078, 2007.

[Lin01] Lin, Y.-B. and Chlamtac, I., *Wireless and Mobile Network Architectures*. John Wiley & Sons, Ltd., Chichester, UK, 2001.

[Lin06] Lin, P., Lin, Y.-B., Yen, C.S. and Jeng, J.-Y., Credit allocation for UMTS prepaid service, *IEEE Transactions on Vehicular Technology*, **55**(1): 306–317, 2006.

Appendix F

An Example of an IMS Charging Application Server

This appendix uses the WebLogic SIP server as an example to show how IMS charging is configured in a SIP application server. The WebLogic SIP server is a commercial product that contains a simple API supporting the Diameter protocol, including the Diameter Base Accounting and Diameter Credit Control application described in Chapter 5. To use the Diameter functionality, the WebLogic domain must be properly configured. The configuration procedure consists of the following steps:

- Enable the Diameter custom resource.
- Create a network channel for Diameter.
- Configure the Diameter nodes and applications.

Based on the materials in [Gio07], this appendix shows how to configure and use the Diameter functionality of WebLogic. Similar procedures are used in other Diameter tools.

F.1 Rf/Ro Interface and Session Initialization

Before a deployed application in the WebLogic SIP server can use Diameter Rf (for offline charging) or Ro (for online charging) functionality, it has to obtain the `RfApplication` or the `RoApplication` object, respectively. This can be accomplished with the code illustrated in Figure F.1, assuming that we are in a SIP or a HTTP servlet class.

Charging for Mobile All-IP Telecommunications Yi-Bing Lin and Sok-Ian Sou

```
ServletContext sc = getServletConfig().getServletContext();
Node node = (Node)sc.getAttribute("com.bea.wcp.diameter.Node");
if(node == null) {
    throw new ServletException("Can't get Node. Check diameter.xml");
}
// Rf interface initiation
RfApplication rfApp = (RfApplication)
node.getApplication(Charging.RF_APPLICATION_ID);
if(rfApp == null) {
    throw new ServletException("Can't get RfApplication. Check
                 diameter.xml");
}
// Ro interface initiation
RoApplication roApp = (RoApplication)
node.getApplication(Charging.RO_APPLICATION_ID);

if(roApp == null) {
    throw new ServletException("Can't get RoApplication. Check
                 diameter.xml");
}
```

Figure F.1 WebLogic code for the Rf/Ro interface initiation

Formally, a Diameter session is "a related progression of events devoted to a particular activity", according to RFC 3588 [IET03]. Practically, it begins with ACR (START_RECORD) or CCR (INITIAL_REQUEST) and ends with ACA (STOP_RECORD) or CCA (TERMINATION_REQUEST) described in Chapter 5. In the case of a one-time event, the session consists only of the request and the answer. All messages belonging to one session are correlated by a common value of the Session-Id AVP. In the WebLogic SIP server API, Diameter sessions are mapped to `com.bea.wcp.diameter.Session` objects. The `Session` class handles the Session-Id AVP. It has special subclasses for the Rf and the Ro interfaces, namely `RfSession` and `RoSession`, which simplify handling of requests and answers specific for Diameter charging. The session object can be created using `Rf/RoApplication` as illustrated in Figure F.2.

In addition, Diameter sessions are serializable, and can be stored as attributes in the `SipApplicationSession` object. Through this object, the WebLogic SIP server automatically links the session to the active call state. Upon receipt of a message, the container will automatically retrieve the call state to find the corresponding Diameter session.

```
// Rf session creation
RfApplication rfApp = ...
RfSession rfSes = rfApp.createSession();

// Ro session creation
RoApplication roApp = ...
RoSession roSes = roApp.createSession();
```

Figure F.2 WebLogic code for Rf/Ro session creation

F.2 Creating Rf/Ro Requests

Creating an Rf request is quite simple. We start with a session-based request example. As explained in Section F.1, after initiation of the `RfApplication` object and creation of the `RfSession` object, we use the `RfSession` object to create a new Accounting Request. Since this is the first request, the `requestType` will be assigned the value START:

```
ACR acr = session.createACR(RecordType.START);
```

To create an event request, we write the code

```
ACR acr = session.createACR(RecordType.INTERIM);
```

When creating a new Accounting Request, the AVPs listed in Table F.1 are automatically populated. In this table, `diameter.xml` is the configuration file for the charging application server.

Other AVPs can be added by calling one of the overloaded versions of the `addAvp` method:

Table F.1 The AVP attributes in an accounting request

Attribute	Value
Session-Id	Automatically generated
Origin-Host	According to node settings in `diameter.xml`
Origin-Realm	According to node settings in `diameter.xml`
Acct-Application-Id	3, meaning Diameter-based accounting
Destination-Host	Value of the `cdf.host` parameter of `RfApplication`, as set in `diameter.xml`
Destination-Realm	Value of the `cdf.realm` parameter of `RfApplication`, as set in `diameter.xml`

Note: `diameter.xml` is the configuration file for the charging application server.

```
Acr.addAvp(Attribute.EVENT_TIMESTAMP,new Integer(timestamp));
```

Requests for the Ro interface (e.g., credit control requests) are created in a very similar way as for creating Accounting Requests. An example should clarify everything:

```
CCR ccr = roSes.createCCR(RequestType.INITIAL);
```

Note that request types for credit control are different from record types for Accounting; for example, START vs. INITIAL. An event request can be created directly from `RoApplication` (without explicitly creating a session) as follows:

```
CCR eventCcr = roApp.createEventCCR();
```

In both cases, the WebLogic SIP server automatically sets the AVPs listed in Table F.2.

Table F.2 The AVP attributes in a credit control request

Attribute	Value
Session-Id	According to node settings in `diameter.xml`
Origin-Realm	According to node settings in `diameter.xml`
Auth-Application-Id	4, meaning Diameter Credit-Control
Destination-Host	Value of the `ocs.host` parameter of `RoApplication`, as set in `diameter.xml`
Destination-Realm	Value of the `ocs.realm` parameter of `RoApplication`, as set in `diameter.xml`
CC-Request-Type	As indicated by the argument of `createCCR()`; the value is EVENT_REQUEST (4) in the case of `createEventCCR()`
CC-Request-Number	A number incremented for each CCR created in the session

Note: `diameter.xml` is the configuration file for the charging application server.

Other AVPs can be added by using the same method shown previously. Diameter-base attributes are available as static fields of the `Attribute` class. Additionally, charging-related attributes can be found in the classes `Charging` and `CreditControl`. The WebLogic SIP server does not limit the user to these predefined attributes. New ones can be created by using one of the `Attribute.define()` methods. The `Attribute` class contains predefined constants for all base attributes. An example of how to add an AVP is described as follows:

```
public static final Attribute SUBSCRIPTION_ID_TYPE =
  Attribute.define(666,"Subscription-Id-Type", Type.INTEGER);
```

In the case of adding an AVP already defined by the container or by the user, the WebLogic SIP server throws an exception. Having added all the necessary AVPs, it is finally time to send the CCR. There are two ways to accomplish this. Either by using the instruction

```
ccr.send();
```

or by using the instruction

```
CCA answer = (CCA)ccr.sendAndWait();
```

The second method sends the CCR and blocks execution until an answer is received or a timeout occurs.

F.3 Receiving Answers

Answers can be received in two ways in the WebLogic SIP server Diameter API. The first one is synchronous based on the `Request.sendAndWait()` method. This method sends the request and blocks the calling thread until an answer is received or the request times out. It returns the received answer. The synchronous way of sending requests is appropriate for applications in which blocking the thread is not an issue. The second, asynchronous, way separates the actions of sending and receiving. The request is sent by calling the method `Request.send()`. This method returns immediately without waiting for an answer. When the answer arrives later, a listener will be notified by calling its `rcvMessage()` method. The listener must implement the `SessionListener` interface, and it must be set on the session before the answer is to be received. An example is illustrated in Figures F.3–F.5. Figure F.3 lists the code for a `listener` class.

Figure F.4 shows how to create a session and a listener. Note that both codes in Figures F.4 and F.5 are inserted inside some other (or the same) class, in a method responsible for sending the request.

After the code in Figure F.4 has been executed, a request can be created and sent as illustrated in Figure F.5.

The implementation with the listener also permits requests to be received that are sent by the WebLogic SIP server within the current session (e.g., an Abort Session Request is sent when the server decides to immediately close the session). Requests

```
class MyListener implements SessionListener {
   public void rcvMessage(Message message) {
      System.out.println("Received a message: " + message);
      if(message.getCommand().equals(CreditControl.CCA)) {
         System.out.println("The message is a Credit-Control-Answer");
      }
   }
}
```

Figure F.3 The `listener` class

```
//Create session and listener
RoSession roSes = roApp.createSession();
MyListener myListener = new MyListener();

//Set the listener on the session
roSes.setListener(myListener);
```

Figure F.4 Creation of session and listener

```
CCR ccr = roSes.createCCR(RequestType.INITIAL);
ccr.addAvp(...);
ccr.send();
//send() returns immediately. Answer will be received in
//myListener.rcvMessage()
```

Figure F.5 Creating and sending a request

are delivered to the listener's `rcvMessage()` in the same way that answers are. It is the application's responsibility to provide different behaviors for requests and answers. The WebLogic SIP server automatically detects a timeout condition when no answer is received for a request during a configured time. The default timeout value is set in the file `diameter.xml`. It may also be specified on a per-request basis, passing an argument to the `send()` or the `sendAndWait()` methods. An example of charging application JAVA code based on the above description can be found in [Gio07].

F.4 Error/Timeout Handling and Debugging

The WebLogic SIP server handles the timeout by creating a response with result code DIAMETER_UNABLE_TO_DELIVER. The response is never actually sent over the network, but it is passed to the application that sent the request. This way of handling timeouts is analogous to the one used in SIP. Also, the WebLogic SIP server can throw one of the following exceptions:

- `MessageException` is thrown for protocol errors, such as an invalid message format.
- `AvpException` is thrown for invalid and/or missing AVP.

The WebLogic SIP server has logging facilities that can be used to trace incoming and outgoing messages. Message debugging can be configured by using the console. In addition, message debugging can be set with the option

```
Ddiameter.Debug=true
```

in the script file (an `sh` file if running on a Unix-like machine, or `cmd` on the Windows platform). Debug messages are sent directly to the console. Besides the debug settings in the WebLogic SIP server, it is also helpful to use a network sniffer. This type of program shows packets as they travel through the network. Probably the most popular one is the free open source Wireshark (formerly called Ethereal). For more information and demonstrations of the WebLogic Diameter, readers are referred to [Gio07]. The material for this appendix is provided by S. Gioia and T. Radziszewski.

References

[Gio07] Gioia, S. and Radziszewski, T., Understanding the IMS charging architecture, http://dev2dev.bea.com/pub/a/2007/07/IMC-charging-architecture.html, 2007.

[IET03] IETF, Diameter Base Protocol. IETF RFC 3588, 2003.

Appendix G

Non-IP-based Prepaid Phone Service

This appendix describes the non-IP-based prepaid phone service by reiterating Chapter 17 of [Lin01]. The Prepaid phone service is a telecommunications service that requires a customer to pay in advance before making calls. Traditionally, coins played an important role in the prepaid phone service, until the telephone companies realized that coins presented a range of problems. For one, extra overhead was incurred from periodic collection of coins. For another, coin payphones were more likely to be damaged by vandalism. To avoid equipment damage and revenue loss, prepaid cards were invented (see Section 3.2.1), and have become the fastest-growing payment method. The average availability of so-called smart card-based payphones is more than 95%, while the comparable figure for coin phones is less than 70%. Prepaid telecommunications services were offered in Europe and Asia in 1982; they became popular in the United States in 1992. Since then, the mobile prepaid service has been growing exponentially all over the world. Asian countries such as the Philippines, Australia, Hong Kong, Singapore, and Taiwan have already experienced success with prepaid services.

The opposite of prepaid service is postpaid service, whereby the customers pay the telecommunications services after a period of time, typically a month. Postpaid services are limited by (i) the high deposit required when setting up service, and (ii) the risk of bad debt. These disadvantages can be removed or reduced by using prepaid services which allow a smaller prepayment that immediately goes toward the customer's usage.

Initially, prepaid cards used in payphone applications were simply token cards whose main benefit was to address the theft and vandalism issues associated with the use of cash. Later, more advantages have been exploited. From the service provider's viewpoint, a prepaid service can significantly reduce the business operation costs.

Charging for Mobile All-IP Telecommunications Yi-Bing Lin and Sok-Ian Sou

Because no service is provided if the end users do not deposit enough money in the accounts, the additional costs of credit checking and collection departments can be eliminated. In other words, a service can be offered to people with bad credit (which can be as high as 40% of the prepaid customer population); typically, revenue is received one and a half months earlier, compared with the postpaid service. And, because it is not necessary to bill prepaid subscribers, printing invoices or managing accounts can be avoided.

From the customer's perspective, prepaid services provide immediate service without the need for long-term contracts, allowing better control of spending. Particularly, many end users (especially the young) just want to enjoy the service; they do not want to fill in detailed subscription forms. Their needs can be satisfied by a prepaid service. Imagine buying a prepaid GSM subscription in the supermarket! Furthermore, prepaid service eliminates a monthly subscription charge and reduces the perceived risk of stolen or lost cards. The preceding discussion for prepaid telephone services implies that the customer-imposed barrier to entry is relatively low. This conclusion is particularly true for a mobile phone service. FarEasTone, a major Taiwan mobile operator, reported that in May 1999, more than 40% of its customers subscribed to the prepaid service in one year after its launch in mid-1998.

Although the fixed and the mobile prepaid services share many characteristics, they have two major differences: First, fixed prepaid service allows outgoing calls only, whereas mobile prepaid service allows both incoming and outgoing calls; second, a fixed telephone service provider knows nothing about the prepaid customers. Thus, no account management is required in the fixed prepaid service. A subscriber simply buys a calling card and starts making calls. As soon as the prepaid balance is exhausted, the card becomes inactive. To provide mobile prepaid service, a prepaid service center is required to perform account management and other functions, as we will discuss later.

This appendix describes and compares four non-IP-based mobile prepaid service solutions. We first identify the requirements for mobile prepaid service. Then we describe mobile prepaid service approaches, including wireless intelligent network, service node, hot billing, and handset-based. These approaches are analyzed and compared.

G.1 Non-IP-based Mobile Prepaid Services

We use GSM as an example to illustrate how the non-IP-based prepaid service works. In GSM prepaid service, a customer subscribes to the GSM service with a prepaid credit. This credit is either coded into the *Subscriber Identity Module* (SIM) card or kept in the network. In many service areas, initialization of a prepaid customer must be completed within a certain number of days after subscription. In Taiwan, prepaid service is available within one hour after purchasing the service. Whenever the customer originates a prepaid call, the corresponding payment is decremented from the prepaid credit. A status report showing the credit balance can be obtained from the SIM card or the network.

If the balance is depleted, the customer cannot originate calls, but may be allowed to receive phone calls for a predetermined period (e.g., six months). To recover the prepaid service, the balance has to be recharged by purchasing a top-up card, which is similar to a lottery scratch card. When the seal is scratched off, a secret code appears. The customer dials a toll-free number and follows the instructions of an *Interactive Voice Response* (IVR) to input the *Mobile Station Integrated Services Digital Number* (MSISDN) and the secret code. The system verifies and refreshes the account if it is a valid code. If, on the other hand, the prepaid balance is not depleted at the end of a valid period, the balance is automatically reset to zero. After a designated period of time, the unused prepaid credit may be considered abandoned, and becomes the service provider's (or government's) property.

Several mechanisms in the mobile prepaid service are not found in the fixed prepaid service:

- An extra billing system for mobile prepaid service is required. Various rate plans must be maintained based on destination of call (local, national, or international), particular numbers (premium rate or free phone number), partitioning of airtime versus land network usage, call forwarding charges, and so on. Tariff switching is required when a customer moves among different service areas during the prepaid calls.
- A real-time usage metering function must be built into the prepaid service system to monitor the amount of remaining credit on the customer account. This function measures the services provided to the customer and decrements the balance during the service or immediately after the service is completed.
- Sales taxes are generally collected at the *Point of Sale* (POS) for prepaid service. Other taxes (universal service fees, relay service fees, presubscribed line charges, and federal access charges) are embedded in the cost of the prepaid product, and then allocated by the service provider accordingly. Due to the mobility of prepaid customers, mobile service providers must understand that usage originates from the various tax jurisdictions based on mobility databases such as HLR and VLR.
- A customer care mechanism maintains items including customer activation and deactivation times, credit value, remaining time period, PIN information, deletion time, reason for subscriber deletion, and so on. An easy credit-refresh mechanism is essential to encourage customers to continue the prepaid service. The mechanism should also generate solicited responses to customer balance inquiries, and unsolicited warnings when the customer's remaining balance drops below a predetermined threshold.

Four solutions have been proposed to implement non-IP-based prepaid services [And98, Dav98, Yan99]. In the remainder of this appendix, we describe these service solutions and show how they implement the prepaid mechanisms just discussed.

G.2 Wireless Intelligent Network Approach

The *Wireless Intelligent Network* (WIN) approach is considered a complete solution to the traditional prepaid service, which has evolved into the prepaid solutions for the all-IP network described in Chapter 5. In this approach, a *Prepaid Service Control Point* (P-SCP) communicates with the MSC through an intelligent network protocol over the SS7 signaling network. This protocol follows the same concept as the IN protocol described in Section 5.1. Several WIN triggers are defined. At prepaid call setup, and during the call-holding period, the MSC encounters WIN triggers at different stages, which remotely instruct the P-SCP to carry out decisions about how that call should be processed based on the prepaid applications. All billing information for a prepaid customer is stored in the P-SCP. The mobile network may need extra SS7 links to accommodate signaling traffic generated by the WIN prepaid mechanism.

G.2.1 WIN Call Origination

Figure G.1 illustrates the WIN call origination, which comprises the following steps:

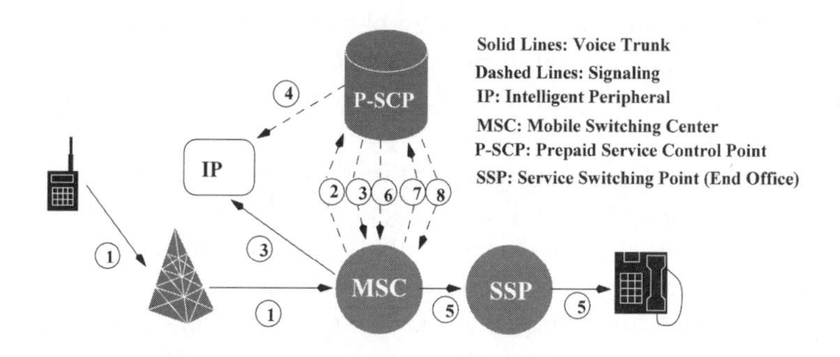

Figure G.1 WIN prepaid cell origination

Step 1. The prepaid customer (the MS) initiates a call by dialing the called party's telephone number.

Step 2. The MSC encounters the WIN call-setup trigger. The call-setup process is suspended, and a prepaid call request message is sent to the P-SCP. The message includes the MSISDN, location information of the MS, and the called party's telephone number. The P-SCP determines whether or not the customer can make the call by querying its database. Based upon threshold processing parameters defined in the prepaid billing system, the P-SCP may deny or accept the call. If the call is accepted, Step 3 is executed.

Step 3. The P-SCP instructs the MSC to establish an ISDN (voice) link to the intelligent peripheral.

Step 4. The P-SCP instructs the intelligent peripheral to provide account status notification, such as the account balance and the charging rate for the call to be made to the prepaid customer.

Step 5. The P-SCP asks the MSC to resume the call-setup procedure, and the call is eventually connected. The P-SCP starts a countdown timer. The amount of credit decrement (from the current balance) is derived from carrier-defined threshold parameters, the rate plan, the destination, and time/date dependency.

Step 6. The call terminates when either the balance depletes or the call completes. If the countdown timer ends before the customer terminates the call, the P-SCP instructs the MSC to terminate the call. For normal call completion, this step will not be executed.

Step 7. Once the call is terminated, the MSC encounters a WIN call-release trigger, which sends a disconnect message to the P-SCP indicating the completion time of the call.

Step 8. The P-SCP rates the completed call and updates the customer's prepaid balance accordingly. Then it sends the current balance and cost of the call to the MSC. The MSC releases the call.

In the above procedure, Steps 3 and 4 are optional.

G.2.2 WIN Call Termination

For calling-party-pays billing (which is used in Taiwan), call termination to a prepaid customer is exactly the same as that for postpaid call termination. For both-parties-pay billing (which is used in the United States), a mobile user pays for the air usage and mobility of incoming and outgoing calls. In this case, the message flow of a WIN prepaid call termination is illustrated in Figure G.2, and described in the following steps:

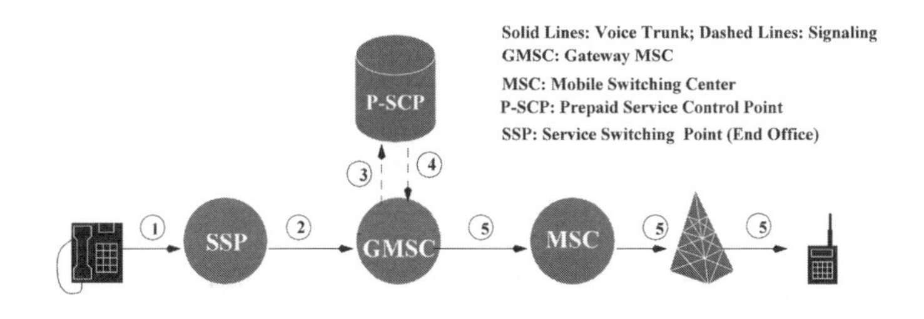

Figure G.2 Prepaid call termination in IN

Step 1. The calling party dials the prepaid customer's MSISDN.
Step 2. The call is forwarded to the *Gateway MSC* (GMSC) of the prepaid MSISDN.
Step 3. The GMSC encounters the WIN call-setup trigger. The call-setup process is suspended and a prepaid call request message is sent to the P-SCP. The message includes the MSISDN and related information.
Step 4. The P-SCP determines whether or not the prepaid customer is eligible to receive the call. Assuming that the call is accepted, the P-SCP instructs the GMSC to resume the call setup procedure.
Step 5. Following the GSM standard *Mobile Station Roaming Number* (MSRN) retrieval and call-setup procedures described in Section 6.1, the call is eventually connected.

In the call, the P-SCP monitors the prepaid customer's balance as described in Step 5 of the prepaid call origination procedure. For both-parties-pay billing, the call-release procedure for prepaid call termination is exactly the same as that for prepaid call origination.

G.2.3 WIN Prepaid Recharging

The message flow of WIN prepaid recharging is illustrated in Figure G.3; the steps are as follows:

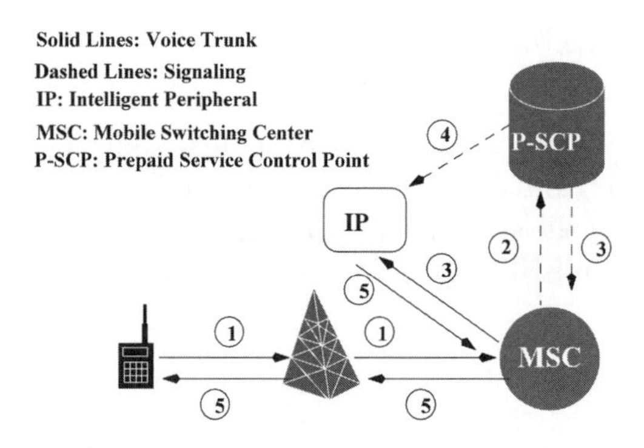

Figure G.3 WIN prepaid recharging

Step 1. A prepaid customer initiates the recharging procedure by dialing a designated number.
Step 2. The MSC encounters the WIN trigger, and a query message is sent to the P-SCP. The message includes the MSISDN of the prepaid phone and related information.

Step 3. The P-SCP instructs the MSC to establish a voice channel to the intelligent peripheral.

Step 4. The P-SCP interacts with the intelligent peripheral to (i) play an announcement and (ii) ask the prepaid customer to enter her PIN number and related information for recharging. Then the P-SCP checks to determine if the voucher is valid.

Step 5. After credit updating, the P-SCP asks the intelligent peripheral to play a new balance announcement. Then it instructs the MSC to disconnect the intelligent peripheral. The MSC releases the call, and the recharging procedure completes.

G.3 Service Node Approach

By 2003, the service node approach was the most widely deployed prepaid service solution at the initial stage of prepaid service provisioning. Many major switching infrastructure providers have predefined call models within their switching architectures. To deploy the prepaid service without interrupting the existing call models, mobile service providers may implement service nodes in their networks to externally control prepaid billing.

As shown in Figure G.4, a service node typically co-locates with an MSC, and is connected to the MSC using standard T1/E1 trunks assigned to a particular block of prepaid numbers. To improve the efficiency of the call-setup procedure, high-speed trunks may be considered for connection.

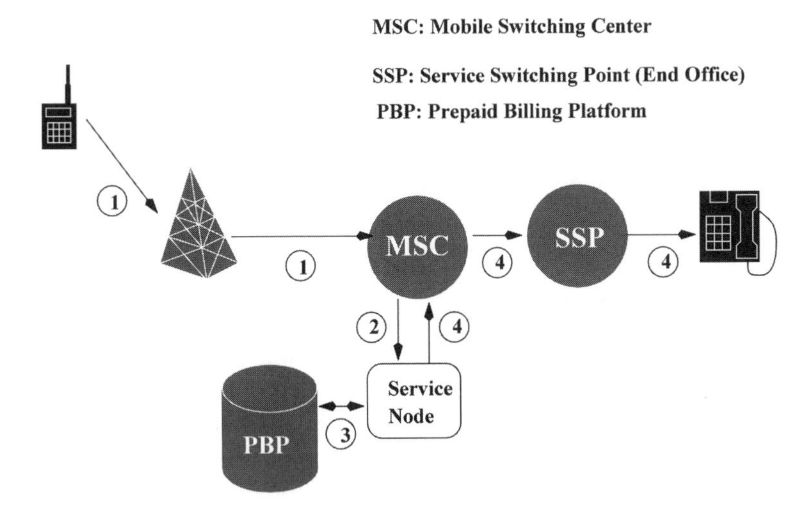

Figure G.4 Service node prepaid call origination

The service node can be implemented by using *Computer Telephony Integration* (CTI) techniques or PC-controlled *Private Branch Exchange* (PBX) techniques. The idea behind CTI is to utilize computer intelligence to manage telephone calls. With *Application Programming Interfaces* (APIs) such as *Telephony API* (TAPI) and *Telephone Services API* (TSAPI), the prepaid applications for the service node can be developed for small installations (e.g., several hundreds of lines). In a PC-controlled PBX, the software (typically written in a high-level language such as C++) in the call control layer can be modified to implement various telecommunications applications. The same platform can also be used to implement the prepaid service node effectively. PC-controlled PBXs provide larger and more cost-effective solutions (in terms of telephone line capacity) than CTI switching. On the other hand, CTI platforms support general APIs that allow fast deployment compared with the PC-controlled PBX platform. The prepaid call origination based on the service node approach is illustrated in Figure G.4; it comprises the following steps:

Step 1. The prepaid customer initiates a call by dialing the called party's telephone number.

Step 2. The MSC identifies that the caller is a prepaid customer. The MSC sets up the trunk to the service node.

Step 3. The service node authorizes the call request by consulting the prepaid billing platform.

Step 4. If the call request is granted, the service node sets up a trunk back to the MSC, and the trunk is connected to the called party from the MSC to the destination SSP following the standard SS7 call-setup procedure. The service node starts credit decrement.

It may be argued that at Step 4, the service node should set up the call directly to the *Public Switched Telephone Network* (PSTN) without passing through the MSC again. By doing so, two ports in the MSC are saved. Typically, this alternative is not considered due to the extra overhead incurred for interworking to the PSTN. In general, small switches such as service nodes are not allowed to connect to the PSTN *Point of Interface* (POI).

G.4 Hot Billing Approach

Hot billing uses CDRs to process prepaid usage. The prepaid CDR is created in the MSC. The information in a CDR includes the prepaid service type, date/time of usage, user identification, the destination of the call, and location information. These records are generated when the calls are completed, and are transported from the MSC to the prepaid service center. The balance of the customer's account is decremented according to the CDRs. When a customer uses up the prepaid credit, the HLR and the *Authentication Center* (AuC) are notified to prevent further service access; the prepaid service center instructs the network to route the next prepaid call attempt to an IVR to play an announcement indicating that the balance has been depleted. The IVR can also communicate with the customer to replenish the prepaid credit by using a top-up card, a credit/debit card, or a credit transfer from a bank account.

Figure G.5 illustrates the interfaces that may be used in the hot billing architecture. In this architecture, a CDR is sent from the MSC to the prepaid service center using protocols such as the *Common Management Information Service Element* (CMISE). This protocol can also be used for communication between the prepaid service center and the HLR. The HLR communicates with the MSC by invoking GSM MAP service primitives similar to those described in Section 4.1. The IVR generates automatic messages that allow the customer accounts to be queried and reloaded. The voice trunks between the IVR and the MSC are set up by SS7 *ISDN User Part* (ISUP) messages.

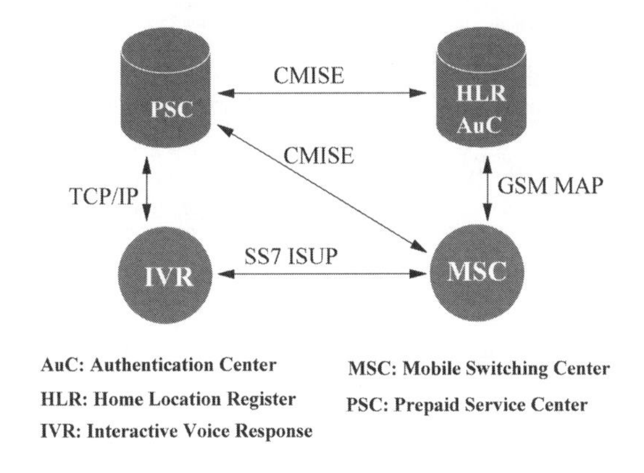

Figure G.5 Hot billing architecture and interfaces

Hot billing depends on real-time data collectors/routers to transport the CDRs from the MSC to the prepaid service center. The HLR/AuC must be updated to allow/prevent prepaid customer access.

G.4.1 Hot Billing Initialization and Call Origination

This subsection discusses service initialization and call origination for the hot billing approach. Note that other prepaid service approaches share a similar service initialization procedure described in the following steps:

Step 1. The customer subscribes to the prepaid service center at the POS or by calling the customer care center.

Step 2. The prepaid service center creates a subscriber data record including IMSI, MSISDN, account of credit, period of validity, tariff model, and other authentication-related information.

Step 3. The prepaid service center activates the prepaid account by sending the customer data to the HLR, which then creates a record for the customer.

To remove a customer from the prepaid service, the prepaid service center simply sends a request to the HLR to delete the customer's record.

The hot billing prepaid call origination procedure is illustrated in Figure G.6; it has the following steps:

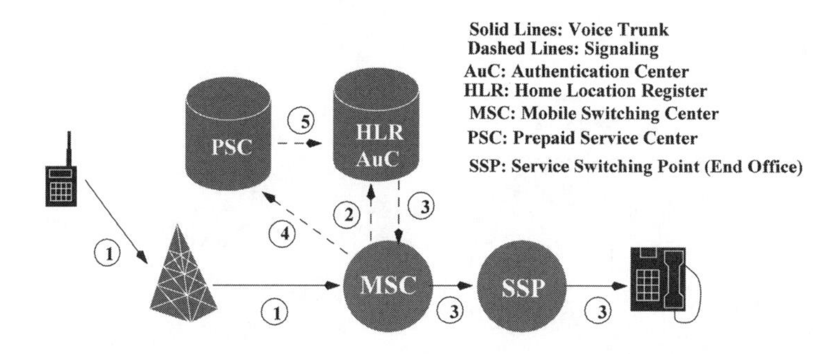

Figure G.6 Hot billing prepaid call origination

Step 1. When a customer originates a prepaid call, the IMSI is sent to the MSC.

Step 2. Based on the IMSI, the MSC instructs the HLR to determine whether or not it is a valid service request.

Step 3. If the verification is successful, the HLR downloads the customer data and a prepaid tag to the MSC. The call is connected.

Step 4. When the call terminates, a CDR is created and sent to the prepaid service center.

Step 5. The prepaid service center decrements the prepaid credit based on the received CDR. If the balance is negative, the prepaid service center instructs the HLR to suspend the prepaid service or to delete the customer's record.

G.4.2 Hot Billing Customer Query and Recharging

Customers can query their current balance through the following steps (see Figure G.7):

Step 1. The customer makes a service query call that, typically, is free of charge.
Step 2. The MSC sends the request, together with the MSISDN of the customer, to the IVR, and sets up a voice path to the IVR.
Step 3. The IVR queries the prepaid service center for account balance information.
Step 4. The IVR plays an announcement to answer the customer.

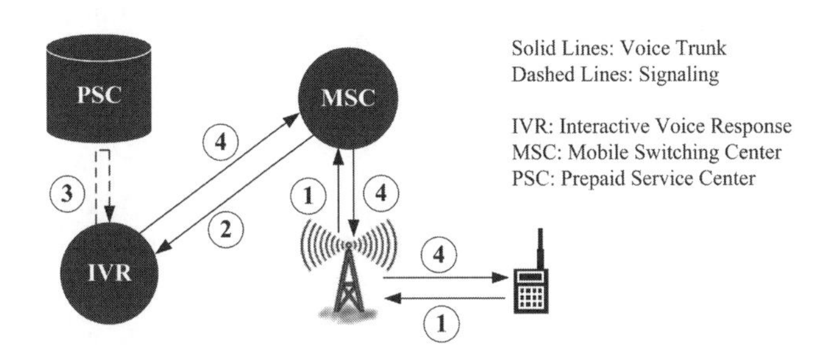

Figure G.7 The hot billing prepaid credit query procedure

When the prepaid credit has been decremented below a threshold, the prepaid service center automatically calls the customer and plays a warning message to remind the customer for credit recharging. The customer may recharge the prepaid account using the top-up card described in Section G.1. This recharging procedure is similar to the credit query procedure illustrated in Figure G.7.

If the prepaid credit depletes during a phone call, the credit becomes negative at the end of the phone call. The negative credit is potentially a bad debt. If the customer does not recharge the credit, it becomes a real bad debt to the service provider. Thus, the "one-call exposure" becomes a major concern of the hot billing approach. A prepaid customer may place the last call and stay connected while the account balance becomes negative. This occurs because most wireless switches do not release the CDR until the call has completed. Some service providers, however, argue that one-call exposure may not be a problem. If the purchased prepaid credits are large enough, the fraudulent user has to exhaust the credit before she can overrun the account, which may not be cost-effective for fraud usage. However, one-call exposure may still be a serious problem, especially when both-parties-pay billing is exercised and a parallel call-forwarding service is available. In this scenario, the MS is used as the call-forwarding mechanism, and the fraudulent user consecutively initiates several calls to the MS in parallel with different forwarding destinations. Consequently, some prepaid solution vendors suggest that call forwarding should not be offered to the prepaid service customers.

To avoid bad debt, the other three approaches described in this appendix decrement the prepaid credit by seconds during a phone call. In the hot billing approach, sending these "real-time" CDRs by seconds to the prepaid service center and processing these

CDRs at the center may incur heavy overhead to the network. Practically, the CDRs are delivered and processed on a per-call basis, and in some cases, on a multiple-call basis. Thus, in the hot billing approach, it is important to select the CDR sending frequency such that the sum of the CDR sending/processing cost and the bad debt is minimized. Also, the service providers may have guarded against one-call exposure by using appropriate call-barring classes. Based on the thresholds under consideration, the network determines when to warn and deny service to a customer.

G.5 Handset-Based Approach

In the handset-based approach, the MS performs credit deduction during the call and determines when the credit limit is reached. In this approach, the prepaid credit is stored in the MS. In the United States, special phones are required for non-GSM systems. For GSM, the credit is stored in the SIM card. We use GSM as an example to describe how a handset-based approach works.

G.5.1 SIM Card Issues

In a typical implementation, the SIM memory is partitioned into two areas. One area stores information such as abbreviated and customized dialing numbers, short messages received, a menu of subscribed services, names of preferred networks to provide service, and so on. Another area stores programs that can be executed to carry out simple commands.

The handset-based approach utilizes the GSM Phase 2 supplementary message *Advice of Charge* (AoC) to transfer the prepaid balance information and the tariff schemes. AoC provides information for the MS to produce a cost estimate of the services used. AoC consists of two service types: *Advice of Charge Charging* (AoCC) and *Advice of Charge Information* (AoCI). To utilize the handset-based prepaid mechanism, the MS must support AoCC. Older MSs that only support AoCI do not work. The supplementary service AoCC is activated for every prepaid customer in an HLR, which will be used in call setup and tariff switching.

Several data fields in a SIM card are used to provide charging information of prepaid service: Accumulated Call Meter (`ACM`), Accumulated Call Meter Maximum (`ACM*`) and Price per Unit and Currency Table (`PUCT`). The `ACM` parameter accumulates the used prepaid units. The `ACM*` parameter records the amount of purchased prepaid credit. When the MS receives the AoC message from the MSC, it converts the AoC into a sequence of SIM commands that modify the SIM data fields (i.e., `ACM`), thereby debiting the customer. The `PUCT` parameter is the value of the home unit in a currency chosen by the subscriber. The value of the `PUCT` can be set by the subscriber and may exceed the value published by the mobile network. The `PUCT` value does not have any impact on the charges raised by the network.

A prepaid service center is required in the handset-based prepaid system, which utilizes the short message service to download executable programs to the SIM card. In call setup and tariff switching, the MSC provides the tariff-charging parameters to the MS, and the MS executes the programs with these parameters for call debiting.

In the SIM card, an extra software filter is required to distinguish prepaid-related short messages from normal short messages. To enhance security, a prepaid-related short message may be authenticated by the SIM card. The GSM specification allows customers to access `ACM` and `ACM*` data fields in the SIM card by using a password, PIN2. To support a prepaid SIM card, PIN2 must be disabled by the card manufacturer when the card is personalized. When the prepaid customer becomes a postpaid customer, PIN2 will be activated by a short message triggered by the subscription switching process.

The SIM Toolkit specification supports proactive commands that enable the SIM card to execute application programs. For GSM Phase 2 SIM cards with larger memory, the cards can run applets downloaded from the SIM Toolkit service. These applets can run security-checking algorithms and simple rating algorithms. These SIM cards can hold tariff table data for various rate plans in their memory buffers.

G.5.2 Handset-Based Call Origination

The prepaid call origination for the handset-based approach is described in the following steps (see Figure G.8):

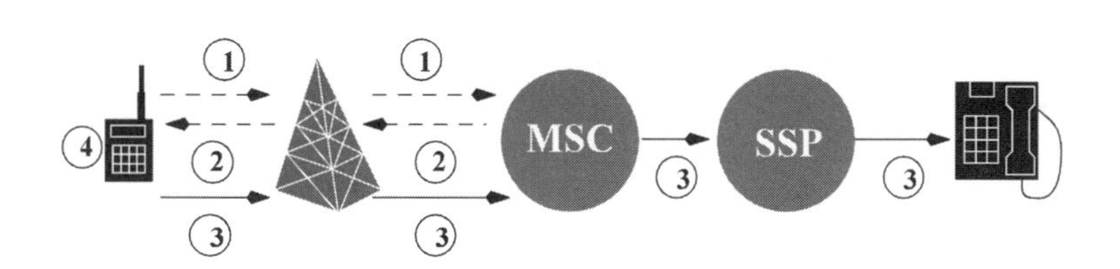

Figure G.8 Prepaid call origination in the handset-based approach

Step 1. The prepaid customer initiates a call by dialing the called party's telephone number.

Step 2. Based on the rate plan and other parameters (such as the destination and time/date dependency), the MSC sends the the AoC e-parameter (including charging information such as `ACM` and `ACM*`) to the MS.

Step 3. If the MS supports AoCC, then it acknowledges the reception of the e-parameters. If this acknowledgment is not received by the MSC, the call is denied. Otherwise, the call is connected.

Step 4. During the call, the MS uses the AoC e-parameters for tariff information. It decrements the credit on the SIM card by incrementing the used credit units in the `ACM`.

If the MS determines that the value of `ACM` has reached that of `ACM*`, the MS disconnects the call and informs the MSC that call release has occurred. This AoC disconnection mechanism works autonomously in the MS without any involvement from the network.

Besides call setup, the AoC e-parameters are transferred to the MS at tariff switching (e.g., billing rate changing, because the MS roams to another area). To reduce the risk of fraud, the handset-based approach may be combined with the hot billing approach. In this case, the prepaid service center in the hot billing approach is included in the prepaid architecture. Figure G.9 shows the message flow of the prepaid call origination for the combined approach; its steps are as follows:

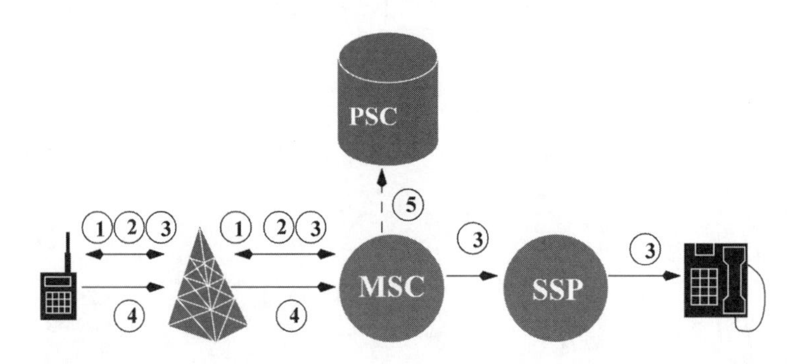

Figure G.9 Prepaid call origination in the combined approach

Steps 1–3. These steps are the same as those for call origination for the handset-based approach. Some of the details (i.e., Steps 2 and 3 in Figure G.6) in the hot billing call setup are not shown here.

Step 4. The call is released when the call completes normally or when the MS notices that the prepaid credit is used up. In either case, the MS sends a message to the MSC for call release.

Step 5. The MSC generates a prepaid CDR, which is then transferred to the prepaid service center. Unlike the regular CDR, the prepaid CDR includes the AoC e-parameters. The prepaid service center updates the prepaid credit as in the hot billing approach. When the `ACM` value is above a threshold, the prepaid service center may automatically send a warning message to the MS for recharging the prepaid credit.

From the viewpoint of the handset-based approach, this combination provides better fraud protection. When the credit in the prepaid service center is different from that in the MS, the service provider may terminate the service for further fraud investigation. From the viewpoint of the hot billing approach, this combination eliminates the possibility of one-call exposure. As soon as the credit is used up, the MS terminates the call, and the situation is reported to the prepaid service center.

Besides the additional implementation complexity, a potential issue for the combined approach is that the charging information (e.g., `ACM` and `ACM*`) may not be consistent for reasons other than fraudulent usage. Thus, synchronization between the prepaid service center and the MS is important.

G.5.3 Handset-Based Prepaid Recharging

As in the hot billing approach, a customer recharges the prepaid credit by purchasing a scratch card, as described in the following steps (see Figure G.10):

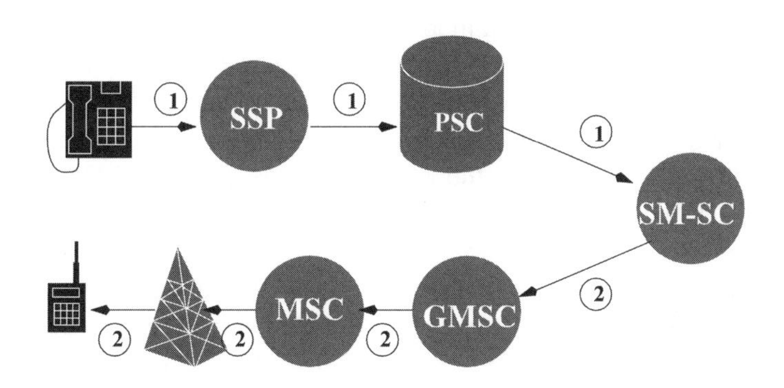

Figure G.10 Prepaid call recharging in the handset-based approach

Step 1. The customer makes a toll-free call that connects to an SSP (with the IVR feature). The prepaid service center validates the secret code (obtained from the scratch card) and the MSISDN of the prepaid customer. If validation is successful, the prepaid service center resets `ACM` to zero; `ACM*` is set to the amount of the new credit.

Step 2. If the recharging procedure is successful, the prepaid service center generates a reload short message. This message is delivered to the MS through the *Short Message Service Center* (SM-SC) in a few minutes, as described in Section 2.2 (see Figure 2.4). If the SM-SC fails to deliver the message (e.g., the MS is turned off), it repeats the message delivery action until it succeeds.

G.6 Comparison of the Prepaid Solutions

Based on the descriptions in the previous sections, we compare the four prepaid service approaches in the following aspects: roaming, scalability, fraud risk, service features, and real-time rating.

G.6.1 Roaming to Other Networks

Assume that the home and the visited systems belong to different service providers. To provide roaming to prepaid customers, an agreement (which can be a part of the roaming agreement) must be made between the home system and the visited system. This agreement is required so that the visited system can (and is willing to) distinguish prepaid calls from the postpaid calls generated by roamers. In most traditional telecommunications scenarios, GSM service providers assign special MSISDN number blocks to prepaid customers. The visited system then identifies a prepaid call based on the MSISDN. There are potential disadvantages to using the MSISDN for prepaid call identification. First, operator number portability [Lin05] will not be allowed. (With number portability, a customer can switch mobile service providers without changing the MSISDN.) Second, service number portability will not be allowed; that is, for the same GSM system, a prepaid customer cannot switch to postpaid service without changing the MSISDN. It seems that identifying prepaid calls by IMSIs will be a better alternative to address these issues. However, the MSC at the visited system may need to be modified so that it can perform prepaid call routing based on IMSIs.

Prepaid charging cannot be performed at the visited system because the home system and the visited system may exercise incompatible prepaid service solutions. Thus, most (if not all) networks require the visited MSC to route the prepaid call back to the home network. This operation is achieved by using the standard “alternate” routing or “optimal” routing that can be easily implemented by setting up routing parameters in the MSC. That is, in the call model of a visited MSC, if the prepaid

MSISDN is recognized, the visited MSC routes the call based on the MSISDN instead of the called party number. Figure G.11 uses the service node approach as an example to illustrate how prepaid call origination is preformed in the visited system.

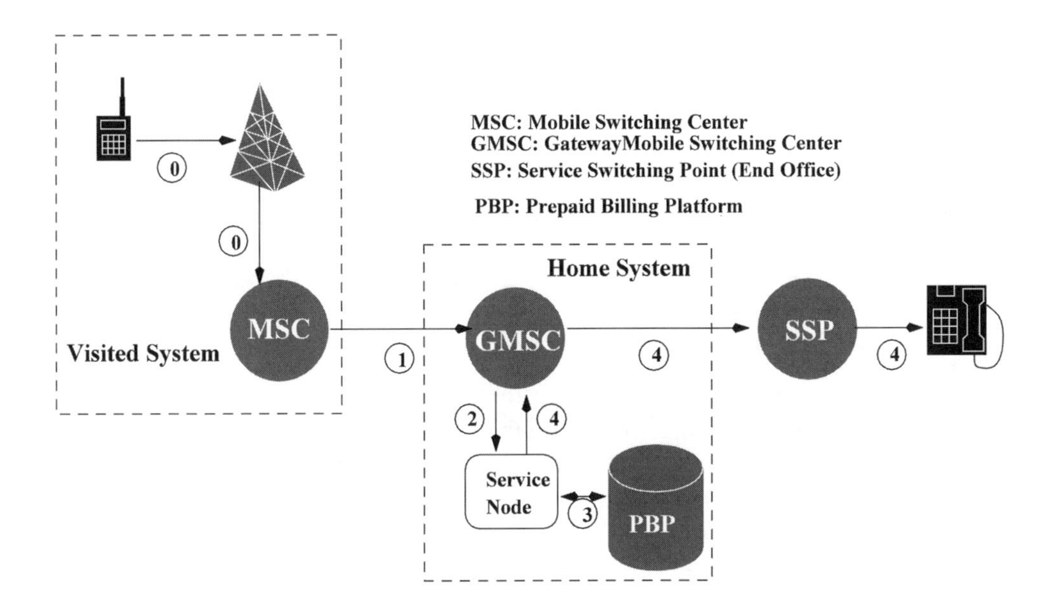

Figure G.11 Prepaid service with roaming to other networks

When the visited MSC receives the prepaid call at Step 0, the MSC routes the call directly to the GMSC of the prepaid MS. The remaining steps are the same as those described in Section G.3. An extra trunk connection is required in this call-setup procedure. Prepaid calls are charged more than postpaid calls, due, in part, to this reason. It is clear that the above procedure is too expensive for international roaming.

G.6.2 Scalability

It is apparent that both the handset-based and the WIN approaches have good scalability. In the hot billing approach, the size of the prepaid customer population is limited to the MSC's ability to process and deliver CDR messages.

In the service node approach, the capacity of the trunks between the service node and the MSC limits the prepaid customer population that can be accommodated in the system. Interestingly enough, in Taiwan, statistics from some service providers exercising the service node approach indicate that GSM network traffic congestion is caused by incoming calls to the prepaid customers, not outgoing prepaid calls. That is, traffic congestion is not caused by the limited capacity of the service node.

G.6.3 Fraud Risk

In the handset-based approach, AoC communication is not encrypted. It is a fairly straightforward exercise to tamper with or ignore AoC by intercepting the debit commands. Also, it is possible to modify the credit illegally in the MS. Thus, it turns out that the handset-based approach has poor fraud protection. Several manufacturers are working on more robust SIM encryption.

The risk of fraud can be high for the hot billing approach due to one-call exposure. As we pointed out before, this occurs because most wireless switches do not release the CDRs until the calls have completed. Fraud risk can be reduced if mid-call CDR sending is exercised. Both the service node and the WIN approaches exhibit low fraud risk.

G.6.4 Initial System Setup

The initial system setup cost and setup time for the handset-based approach is average. This approach does not require changes to the mobile network infrastructure, except that the MSC must support AoC. On the other hand, the prepaid MSs must be GSM Phase 2-compliant to receive AoC messages. Furthermore, special SIM software is required to execute rate plans in the MS. The GSM service providers may be locked to a single-source SIM supplier. This situation is undesirable, especially when the SIM card market is unpredictable.

The initial system setup cost and setup time for the hot billing approach is average. This approach requires the integration of the prepaid service center, the IVR recharging mechanism, and the MSC/HLR.

System setup for the service node approach can be done quickly. The mobile network infrastructure is not modified. The only system setup cost is for the establishment of the service node. For this reason, by the end of 2000, service node was the only working prepaid solution in Taiwan.

The initial system setup time for the WIN approach is long, and the setup cost is very high. In traditional telecommunications, this approach does not provide a fully developed model for other mobile networks to follow. The design of services and switch software development for the intelligent network is complex. For small- and medium-sized service providers, a full implementation of an IN architecture may not be a realistic option. These issues will be resolved when the next generation All-IP networks are introduced.

G.6.5 Service Features

The handset-based approach supports limited service features because the number of rate plans that can be stored in the SIM card is limited and cannot be conveniently updated. The SIM card also relies on the MSC to provide the tariff charging model.

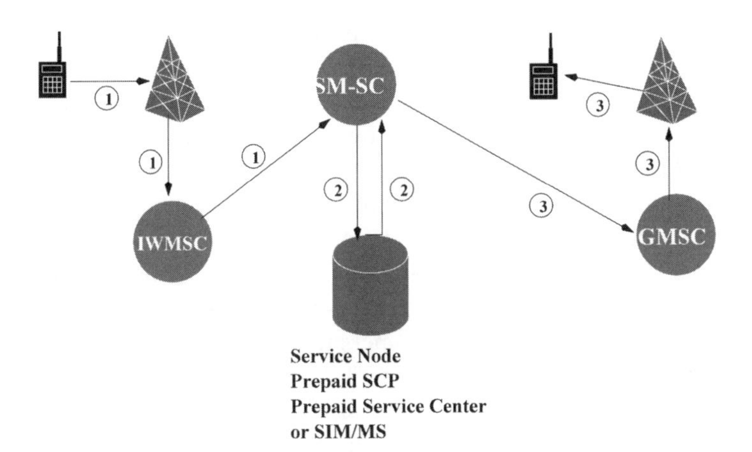

Figure G.12 Prepaid short message service

The service feature provisioning for the hot billing approach is better than average, which is typically limited by the per-post call charging mechanism.

Both the service node and the WIN approaches support flexible service features. For WIN, many service features can be integrated with the prepaid service under the intelligent network platform. None of the four approaches can support the prepaid short message service, because the short message is delivered by the SS7 *Transaction Capabilities Application Part* (TCAP), which cannot be identified by the MSC for charging purposes. Instead, charging for the short message is done at the SM-SC. One solution is to modify SM-SC, such that before delivering a prepaid short message (see (1) in Figure G.12), the SM-SC sends a charging message to MS, PSC, or P-SCP (depending on the approach exercised) for decrementing the prepaid credit of the customer. If the SM-SC receives a positive response (see (2) in Figure G.12), the short message is delivered to the destination (see (3) in Figure G.12).

G.6.6 Real-Time Rating

In the handset-based approach, real-time rating is performed at the MS. In the service node approach, real-time rating is performed at the service node. In the WIN approach, real-time rating is performed at the P-SCP. The hot billing approach cannot support real-time rating. The credit information update depends on the MSC capability of sending CDRs.

G.7 Business Issues

Besides the technical discussion for the prepaid service approaches, we would like to point out that the business case for packaging prepaid services is very important. A few years ago, the prepaid service was over-promoted in Mexico, resulting in the loss of postpaid revenues because the prepaid service was introduced with more favorable rates. Thus, it is important for service providers to balance their prepaid and postpaid marketing and distribution strategies through segmentation (e.g., targeting prepaid service for high-risk customers, or packaging the service as gifts). As mentioned in Section G.1, although service providers may know why customers purchase prepaid services, our experience indicates that they may not know why customers stop using prepaid services, or why customers switch to other service providers. In Taiwan, the prepaid service is considered to be a niche market, and the strategy is to promote program loyalty to convert prepaid customers into postpaid customers.

As a final remark, credit transfer for the mobile prepaid service is clearly an application for electronic commerce, which transfers electronic cash over digital networks in real time. Prepaid billing systems are beginning to influence existing billing systems, where the payment-processing components will need to be tailored for the online nature of the new medium, which is the main focus of this book.

G.8 Concluding Remarks

This appendix described four traditional mobile prepaid service approaches and compared their weaknesses and strengths. Among the four approaches, the handset-based approach is a low-cost, high-risk solution. The wireless intelligent network approach is a high-cost, low-risk solution. The service node approach is a quick, but dirty solution, which allows rapid deployment with limited capacity. The hot billing approach is an average solution that cannot provide real-time service. We note that except for the wireless intelligent network approach, the other three approaches will become outdated, and will not be used in the all-IP networks.

The intelligent network approach is considered as a complete solution for the prepaid service. In this approach, the service control and service development functions are migrated from the MSCs to the *Prepaid Service Control Point* (P-SCP). The SCP contains *Service Logic Programs* (SLPs) and associated data to provide IN services [Uyl98]. When an MSC encounters a prepaid call, it communicates with the P-SCP through SS7 links, asking the P-SCP to decide how the call should be processed. The P-SCP performs the service control functions (e.g., checking credit and activating a countdown timer) based on the customer's credit and sends a response message back to the MSC. After receiving the message, the MSC performs the P-SCP instructions to accept or reject the prepaid call. Since the P-SCP is not on the voice path, the intelligent network solution allows real-time call control with low capacity expansion cost. However, not all traditional carriers are interested in implementing this approach because of the investment in P-SCP and the necessity for software modifications in all MSCs.

For further reading, the CTI technologies that can be utilized for implementing service node prepaid approach can be found in [Fle96]. For the PBX-based service node technologies, the reader is referred to [Lin00]. Details of TAPI and TSAPI are elaborated in [Mic96] and [Cro96, Nov96], respectively. The CMISE standard is specified in [ETS93]. The tax jurisdictions based on mobility databases are discussed in [Sla98].

Review Questions

1. What is a prepaid telephone service? What is a postpaid telephone service? What are the advantages of prepaid service over postpaid service? Do you think a fixed prepaid service can replace a fixed postpaid service?
2. What are the differences between a fixed prepaid service and a mobile prepaid service?
3. What are the four mobile prepaid service solutions? What are the advantages and disadvantages of these solutions?
4. Propose a solution so that for a roaming mobile prepaid call origination, the trunk is not required to set up back to the home system of the prepaid customer. Can your solution apply to international roaming?
5. When a mobile user unsubscribes to the mobile service, the mobile identification number assigned to the user must be reclaimed. This reclaimed number must be kept unused for a period, in a process called number ageing. Which service needs a longer ageing period, prepaid or postpaid? Which country, the United States or Taiwan, needs a longer ageing period?
6. In this appendix, it was stated that traffic congestion is not caused by the limited capacity of the service node in Taiwan. Why? (Hint: Calling-party-pays billing is exercised in Taiwan.)
7. Show the call-setup procedure where both the calling party and the called party are prepaid users, and the called-party-pays policy is exercised.
8. In early 2000, the U.S. cellular carriers were expected to invest billions of dollars to implement cellular billing and for customer care. Two of the most desirable attributes of the cellular billing systems are the flexibility of upgrade and the ability to inform the billing experts quickly about the status of the system, to minimize any possible fraud and improve customer service. To speed up the billing information transmission, a cellular billing transmission standard called

EIA/TIA IS-124 was developed by working group 4 of TIA's TR 45.2 committee. IS-124 allows real-time billing information exchange, which will help control fraud by reducing the lag time created by the use of overnighted tape messages. Version A of IS-124 also accommodates both U.S. AMPS and international GSM carriers, which is desirable for heterogeneous mobile system integration and third-generation core network interworking. An important performance issue of cellular billing information transmission is the frequency of the billing information exchange. In the ideal case, records would be transmitted for every phone call to achieve the real-time operation. In this case, the billing mechanism is the same as the one used in the hot billing prepaid service. However, real-time transmission would significantly increase the cellular signaling traffic and seriously overload the signaling network of the PSTN. In order to achieve quick billing status report, a trade-off is therefore needed between the frequency of the billing information transmission and the signaling traffic. Design a heuristic algorithm to determine the frequency of the billing information transmission. (Hint: See the discussion in [Fan99].)

References

[And98] Arteta, A., Prepaid billing technologies – which one is for you? *Billing World*, pp. 54–61, February, 1998.

[Dav98] Dave, E., Pay now, call later, *Telecommunications*, March, 1998.

[Cro96] Cronin, P., An introduction to TSAPI and network telephony, *IEEE Communications*, pp. 48–54, April, 1996.

[ETS93] ETSI, European Digital cellular telecommunications system (phase 2); common aspects of PLMN network management, ETSI/TC Recommendation GSM 12.01, 1993.

[Fan99] Fang, Y., Chlamtac, I. and Lin, Y.-B., Billing strategies and performance analysis for PCS networks, *IEEE Transactions on Vehicular Technology*, **48**(2): 638–651, 1999.

[Fle96] Flegg, R., Computer telephony architecture: MVIP, H-MVIP, and SCbus, *IEEE Communications*, pp. 60–64, April, 1996.

[Lin00] Lin, Y.-B., PBX-based mobility manager for wireless local loop, *International Journal of Communication Systems*, **13**(3): 185–316, 2000.

[Lin01] Lin, Y.-B. and Chlamtac, I., *Wireless and Mobile Network Architectures*. John Wiley & Sons, Ltd., Chichester, UK, 2001.

[Lin05] Lin, Y.-B. and Pang, A.-C., *Wireless and Mobile All-IP Networks*. John Wiley & Sons, Ltd., Chichester, UK, 2005.

[Mic96] Microsoft, Win32 SDK: Win32 Telephony (TAPI)-MSDN Library, Microsoft, 1996.

[Nov96] Novell Inc. and AT&T Corp., Netware Telephony Service (Telephony Service Application Programming Interface), Release 2.21, Novell Inc. and AT&T Corp., 1996.

[Sla98] Slavick, F., Counting down prepaid services, *Billing World*, pp. 44–50, 1998.

[Uyl98] Black, U., *The Intelligent Network: Customizing Telecommunication Networks and Services*. Prentice-Hall, 1998.

[Yan99] Chang, M.-F., Lin, Y.-B. and Yang, W.-Z., Performance of hot billing mobile prepaid service, *Computer Networks Journal*, **36**(2): 269–290, 2001.

Appendix H

Performance of Service Node Based Mobile Prepaid Service

The service node approach is widely utilized for early-stage deployment of non-all-IP-based prepaid services in the circuit-switched domain and is viewed as a stepping-stone to the intelligent network approach. This approach integrates the functions of the MSC and *Service Control Point* (SCP) in a closed configuration [Pil98]. The service node usually co-locates with an MSC and is connected to the MSC using high-speed T1/E1 trunks. For each prepaid call, the MSC routes the call to the service node for call processing. After the service node performs the service control functions, the prepaid call is routed back to the MSC. Then the MSC sets up a trunk to the switch of the destination. Thus, to set up a mobile prepaid call, it requires two ports on the service node and four ports on the MSC. Since the service node is on the voice path, this approach allows real-time call control. However, the cost of capacity expansion in the service node approach is higher than that of the intelligent network approach. On the other hand, the service node approach can be quickly deployed without upgrading the MSCs that do not support prepaid signaling protocol. This appendix investigates the performance of the service node approach.

H.1 The Service Node Approach

From Section G.3, the service node checks and decrements the prepaid credit periodically during the voice conversation to avoid potential bad debt. However, the capability of the service node may be limited since all service control and call-switching functions are implemented on the service node. Theoretically, by upgrading the processing

Charging for Mobile All-IP Telecommunications Yi-Bing Lin and Sok-Ian Sou

power of the switch or the service node, this approach will permit a real-time credit monitoring. However, in a real mobile phone network operation, the processing budget for a service node should be accurately planned. An example of processing budget planning for a switching equipment can be found in [Lin01]. In commercial operation, a service node may process over 50,000 prepaid calls simultaneously. To support real-time monitoring for so many simultaneous calls, the cost of upgrading processing power is too high, and is not justified. Thus, the operators always ask the following question: "What is the credit checking frequency so that the sum of the credit checking cost and the bad debt is minimized?" This appendix investigates the performance of the service node, and answers the above question. With this knowledge, the operator can choose the appropriate processing power for the service node so that it can support, say, over 50,000 simultaneous prepaid calls at the selected credit checking frequency.

In [Cha02], we proposed an analytic model to derive the expected number of credit checks ($E[N^*_{ch}]$) and the expected bad debt ($E[B^*_L]$) in the service node approach. Let B be the initial prepaid credit of a customer and K be the number of calls that the customer has made when the prepaid credit runs out. Note that in the OCS studies in Appendix B, we use notation X to represent the initial prepaid credit. This appendix uses notation B for the service node based prepaid mechanism. We assume that a customer will consume all prepaid credit before she gives up the prepaid service. When a customer is in a prepaid conversation, the service node periodically decrements the customer's credit by the amount I until either the call completes or the credit becomes negative.

Let $x_i (i = 1, 2, \ldots K - 1)$ be the charge of the ith call. The last (i.e., the Kth) call terminates in one of two cases:

Case (i): the service node detects that the prepaid credit runs out by periodic checks and the call is forced to terminate (see Figure H.1).

Case (ii): the last call completes before the service node detects that the credit is negative (see Figure H.2).

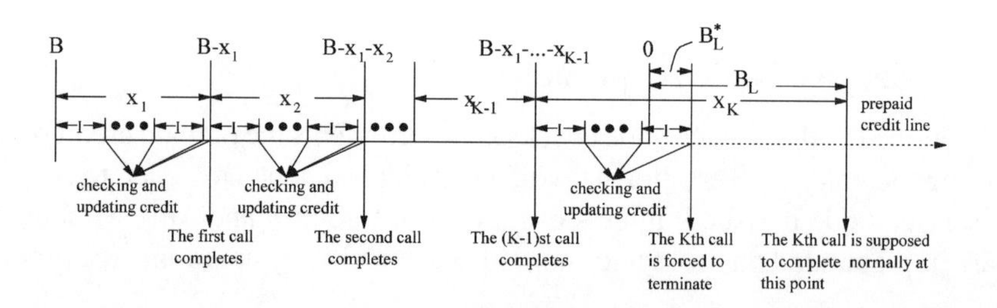

Figure H.1 Case (i): the charges for prepaid calls where the last call is forced to terminate by the service node

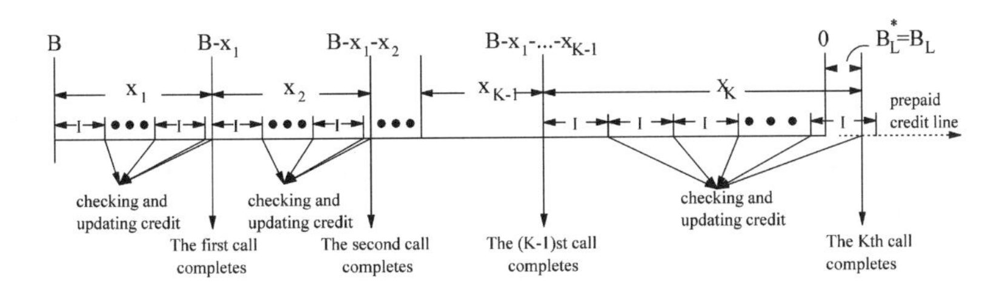

Figure H.2 Case (ii): the charges for prepaid calls where the last call completes before the service node discovers that the credit is negative

In Figures H.1 and H.2, the horizontal line is the "prepaid credit line", which illustrates the decrement of the prepaid credit due to periodic credit checks during the calls (the short vertical lines). The long vertical arrows represent call completions or force-terminations. For the derivation purpose, let x_K be the charge of the last call if the service node does not terminate the call when the credit becomes negative. We assume that $x_i(i = 1, 2, \ldots K - 1)$ are independent and identical random variables and the expected value $E[x_i]$. Let B_L^* be the loss of the service provider and B_L be the corresponding value if the last (force-terminated) call were allowed to complete (i.e., the amount between when the credit is exhausted and when the call completes). Note that B_L^* equals B_L in Case (ii); see Figure H.2.

H.2 Numeric Examples

This section investigates the performance of the service node approach based on the analytic model developed in [Cha02]. Simulation experiments were conducted to validate the analytic results. Each simulation experiment was repeated 500,000 times to ensure stable results. To reflect the situation of the prepaid service in Taiwan, the expected charge of a call is assumed to be NT$36, and the expected prepaid credit *Bs* are NT$100, NT$300, NT$400, and NT$500, respectively.

H.2.1 Effects of the Variation of Call Charges

This subsection studies the effect of the variation of call charges on $E[N_{ch}^*]$ and $E[B_L^*]$ for fixed credit and recharged credit cases. In the fixed credit case, the prepaid credit is a constant. In the recharged credit case, a customer may recharge her prepaid card before the credit runs out. The call charge is assumed to have a Gamma distribution. The Gamma distribution is selected because it has been widely used in mobile telecommunications studies [Sou07a, Sou07b] and can be shaped to represent many distributions. Let C_x be the coefficient of the variation of call charge; C_x equals the ratio of the standard derivation to the mean of the distribution. In our experiment, C_x

ranges from 10^{-3} to 10. A large C_x means that there are more short calls and long calls. In both the fixed credit case (see Figure H.3) and the recharged credit case (see Figure H.4), the service node periodically checks and updates prepaid credit with I = NT\$12. The coefficient of variation C_x ranges from 10^{-3} to 10.

The call charges have a Gamma distribution with the means $E[x_i]$ = NT\$37, NT\$36, and NT\$35, respectively. In the fixed credit case, the prepaid credit equals NT\$500. In the recharged credit case, the amount of the initial prepaid credit is NT\$100, and the amount of each recharged credit is B_r. The number of recharges is assumed to have a geometric distribution and the recharge probability p varies as 1/3, 2/5, 2/3, and 4/5. The mean of prepaid credit $E[B]$ is NT\$500. In this experiment, we only present the results where the call charges have a Gamma distribution with mean $E[x_i]$ = NT\$36. Similar conclusions can be drawn for x_i with various means.

Both figures show that for $C_x < 10^{-1}$, $E[N^*_{ch}]$ and $E[B^*_L]$ are sensitive to $E[x_i]$, but insensitive to C_x. The figures also show that when $C_x > 5$, $E[N^*_{ch}]$ increases sharply in both fixed credit and recharged credit cases. The reason is that when C_x increases, the number of short calls, whose call charges are less than I, also increases. For each short call, a small amount of credit ($x_i < I$) is consumed and a credit check is required. As a result, the number of credit checks increases as the number of short calls increases.

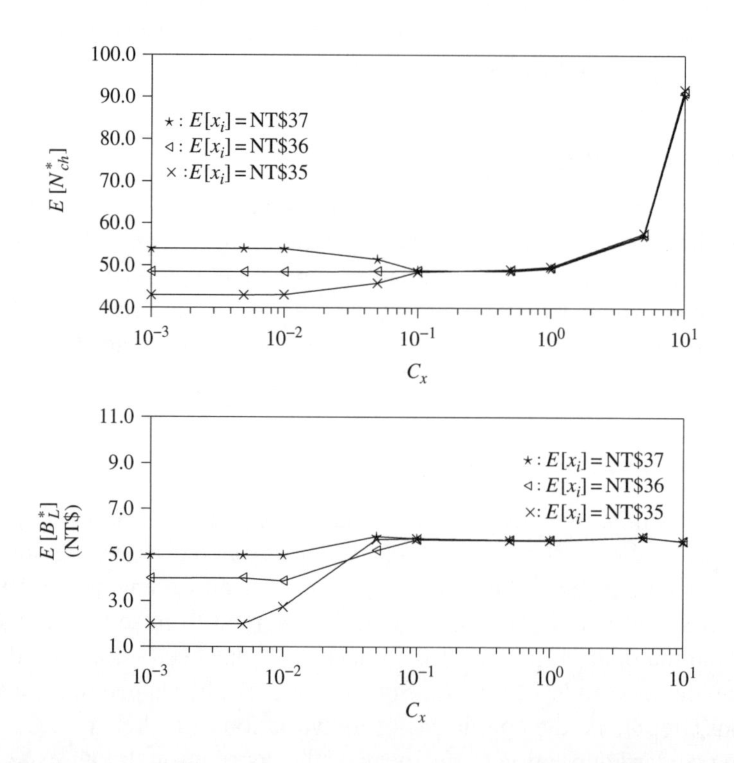

Figure H.3 Effects of C_x in the fixed credit case (B = NT\$500, I = NT\$12)

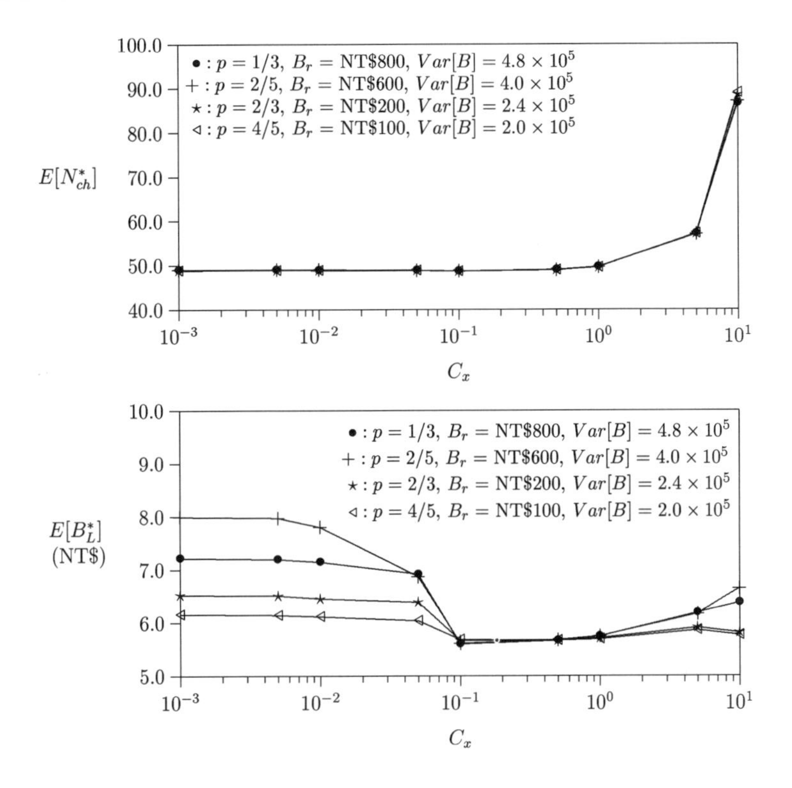

Figure H.4 Effects of C_x in the recharged credit case ($E[B]$ = NT\$500, NT\$100, $E[x_i]$ = NT\$36, I = NT\$12, and the amount of the initial credit is NT\$100)

We call this the short-call effect. To avoid this effect, the prepaid service provider can implement billing policies to discourage short calls (e.g., higher rates in the first minute of a prepaid call and lower rates for the remaining call-holding time).

H.2.2 Effect of I on $E[B^*_L]/I$

Figure H.5 plots $E[B^*_L]/I$ as a function of I in the fixed credit case. The mean $E[x_i]$ of call charge is NT\$36 and I ranges from NT\$0.2 to NT\$36. In this experiment, we consider two scenarios where B = NT\$400 and NT\$500, respectively.

Intuition suggests that $E[B^*_L]$ would be equal to $I/2$. However, Figure H.5 shows that $E[B^*_L]/I$ is equal to 1/2 when $C_x < 1.41$ and I is sufficiently small (e.g., I = NT\$0.2). It is interesting to note that when $C < 1.41$, $E[B^*_L]/I$ almost linearly decreases as I increases. The reason is that as I increases, the probability that the Kth call terminates normally (rather than be forced to terminate by periodical checks) increases. Thus, the expected loss $E[B^*_L]$ becomes smaller than $I/2$. We also observe

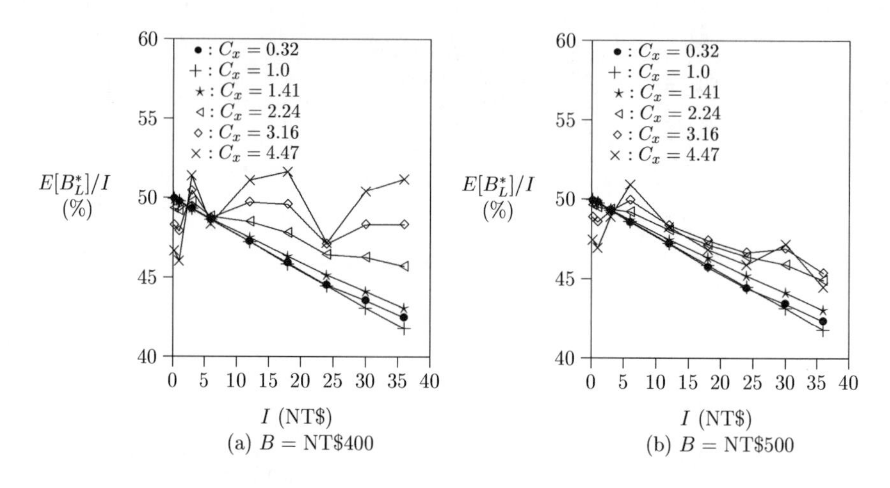

Figure H.5 Effects of I on $E[B_L^*]/I(E[x_i] = \text{NT\$36})$

that when C_x is large (e.g., $C_x > 2.24$), $E[B_L^*]/I$ appears to vary with irregular patterns. As C_x increases, the number of small and large x_i also increases. When C_x is large, the probability that the last call depletes all or most of the credit becomes large. For the same C_x, the bad debt B_L^* depends on the values of B and I. We can see that the patterns of variation in Figure H.5(a) and (b) are different when $C_x > 2.24$.

H.2.3 The Cost Function

Two costs are associated with the service node: the credit checking/updating cost and the bad debt. The credit checking cost and the bad debt are two conflicting factors, since smaller I represents smaller $E[B_L^*]$ and larger $E[N_{ch}^*]$. Consider a cost function $C_s = E[B_L^*] + \phi E[N_{ch}^*]$, where ϕ is the credit checking cost of the service node. The cost C provides the net effect of credit checking cost and bad debt. Figure H.6 plots C_s as a function of ϕ and I, where $E[x_i] = \text{NT\$36}$ and $Var[x_i] = 1296$. Both fixed credit cases (Figure H.6(a) and (b)) and recharged credit cases (Figure H.6(c) and (d)) are considered in this experiment. The triangle mark in the curves represents the cost for the optimal I.

Consider the fixed credit case where B = NT$500. For $\phi = 0.05$, the credit checking cost is high and I = NT$6 should be selected. For $\phi = 0.001$, the credit checking cost is low and I = NT$1 should be selected. In addition, for the same ϕ value, the optimal I value increases as B increases. Although the above results are intuitive, our analysis quantitatively computes the prepaid service overhead to select the optimal checking interval I according to the capability of the service node. For the examples in Figure H.6 (which are consistent with the real network operation), acceptable I values range from NT$1 to NT$6.

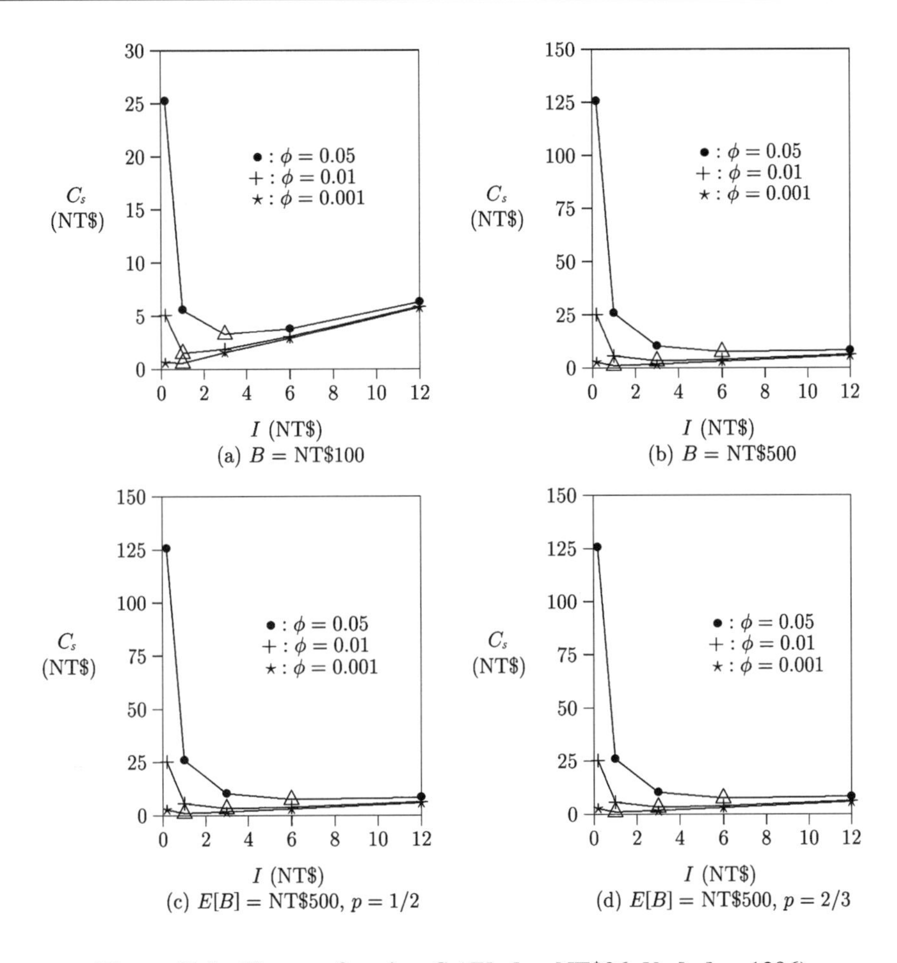

Figure H.6 The cost function $C_s(E[x_i] = \text{NT\$36}, Var[x_i] = 1296)$

H.3 Concluding Remarks

This appendix studied the service node approach for the prepaid service. We investigated the performance in the fixed credit and the recharged credit cases. The following results are observed:

- If the call pattern of a prepaid customer is very irregular (i.e., there are many short calls and many long calls), it is desirable to make more credit checks in the service node. To avoid a large number of credit checks on the service node, the service provider can implement billing policies that discourage short calls (e.g. higher rates in the first minute of the prepaid call and lower rates for the remaining call holding time).

- Intuition suggests that the expected bad debt approximates to half of the amount of one credit check. However, the above observation does not hold when the variation of call charge is high or the amount of single credit check is large.
- A cost function was used to determine the minimal cost for the service node approach. The minimal cost can be achieved by properly setting the credit checking/updating interval to balance against the workload of the service node with the bad debt.

H.4 Notation

- B: the initial prepaid credit
- B_r: the amount of credit for each recharge
- C_s: the cost function of the credit updating and the bad debt
- C_x: the coefficient of the variation of call charge
- $E[N_{ch}^*]$: the expected number of credit checks
- $E[B_L^*]$: the expected bad debt
- I: the amount the service node periodically decrements the customer's credit when a customer is in conversation
- K: the number of calls that a customer has made when the prepaid credit runs out
- p: the recharge probability
- x_i: the charge of the ith call ($i = 1, 2, \ldots K - 1$)

References

[Cha02] Chang, M.-F., Yang, W.-Z. and Lin, Y.-B., Performance of service-node-based mobile prepaid service, *IEEE Transactions on Vehicular Technology*, **51**(3): 597–612, 2002.

[Lin01] Lin, Y.-B. and Chlamtac, I., *Wireless and Mobile Network Architectures*, John Wiley & Sons, Ltd., Chichester, UK, 2001.

[Pil98] Pilcher, R., Intelligent networks move advanced services ahead, *Telecommunications*, November, 1998.

[Sou07a] Sou, S.-I., Lin, Y.-B., Wu, Q. and Jeng, J.-Y., Modeling prepaid application server of VoIP and messaging services for UMTS, *IEEE Transactions on Vehicular Technology*, **56**(3):1434–1441, 2007.

[Sou07b] Sou, S.-I., Hung, H.-N., Lin, Y.-B., Peng, N.-F. and Jeng, J.-Y., Modeling credit reservation procedure for UMTS online charging system, *IEEE Transactions on* Wireless *Communications*, **6**(11): 4129–4135, 2007.

Index

Charging for Mobile All-IP Telecommunications Yi-Bing Lin and Sok-Ian Sou